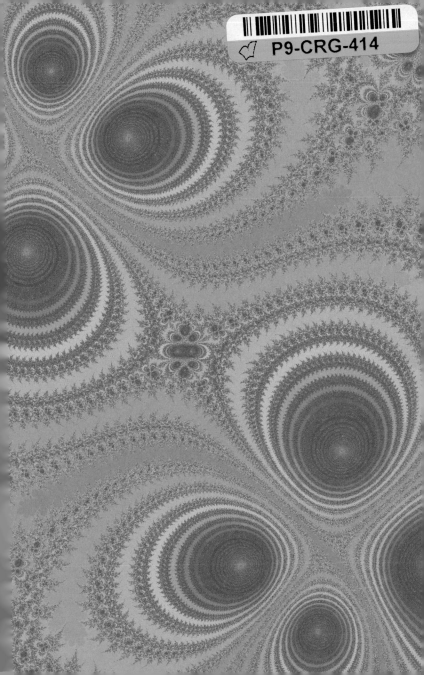

INFINITY
IN YOUR
POCKET

INFINITY
IN YOUR
POCKET

Over 3,000 theorems,
facts, and formulae

MIKE FLYNN

BARNES & NOBLE BOOKS
NEW YORK

This edition published by Barnes & Noble, Inc.,
by arrangement with Elwin Street Limited

2005 Barnes & Noble Books

M 10 9 8 7 6 5 4 3 2 1

ISBN 0-7607-6649-5

Conceived and produced by
Elwin Street Limited
79 St John Street
London EC1M 4NR
www.elwinstreet.com

Editor: Iain Boyd
Designer: Ian Hunt
Illustrations: Richard Burgess
See page 144 for picture credits

Printed in Singapore

Contents

7 **NUMBERS & SHAPES**
The Concept of Numbers 8 Types of Numbers 12
Fractions 16 Decimals 16 Percentages 16 Indices 17
Logarithms 17 Powers of 10 18 Dimension 19
Classification of Two-dimensional Shapes 19 Solid
Figures: Polyhedrons 22 Topology 24 The Beauty of
Numbers 26

29 **USING NUMBERS**
The Development of Measuring Systems 30 SI Units 33
Time Measurement 34 Theory of Measurement 39
Scientific Measurements 39 Simple Graphs and
Charts 41 Probability and Statistics 44

47 **ALGEBRA & TRIGONOMETRY**
Algebra 48 Coordinate Systems 51 Graphs and
Functions 52 Calculus From First Principles 55
Properties and Types of Angles 57 Parts and Properties
of Circles 59 Trigonometry 60

63 **PHYSICS & CHEMISTRY**
Force and Motion 64 Laws of Motion 64 The Nature
of Energy 68 Heat and Temperature 68 Pressure 70
The Gas Laws 71 Pressure and Depth 72 Fluid
Dynamics 72 Electromagnetism 72 Electricity 72
Electric Circuits 74 Electromagnetic Field 74
Elements 75 The Periodic Table 78 Einstein's Theories
of Relativity 80 $E = mc^2$ 82 Radioactivity 83 Fission,
Fusion, and the Bomb 84 Subatomic Particles 86
A Brief History of Cosmology 87 Hubble's Law 88
Big Bang Theory 89 Steady State Theory 90 Brane
Theory 91 Matter Fields 92 Infinity Redefined 92

93 COMPUTERS & DIGITIZATION

Calculating Machines 94 Computer Components 98
Alphanumeric Characters 100 Digital Sound 100
Digital Pictures 101 Digital Video 102 Data
Storage 102 Moore's Law 103 Computers and
Science 104 Virtual Worlds 104 The Internet 106

109 LOGIC, CHAOS THEORY, & FRACTALS

Plato's Influence on Mathematics 110 Euclidean
Mathematics 110 Non-Euclidean Mathematics 110
Physics and Field Equations 111 Chaos Theory 112
Fractals 115

119 GENERAL REFERENCE

Squares and Cubes 120 Roman Numerals 120 Areas
and Volumes of Two- and Three-dimensional Shapes 121
Conversion Tables 122 SI Units and Definitions 124
SI Measurements 125 SI Quantities 126 Logarithm
Table for Sine, Cosine, and Tangent 127 Mathematical
Symbols 130 Basic Rules of Algebra 130 Elements in
the Earth's Crust 131 Common Names and Formulae of
Important Compounds 131 Temperature Scales 132
Melting and Boiling Points of Elements 132 Recent
Discovery of Elements 133 Ions and Radicals 133 Local
Group of Galaxies 134 Brightest Stars 135 Earthquake
Measurements 135 Physics Symbols 136 Physics
Formulae 136 Useful Fractions, Decimals, and
Percentages 138 Metric Prefixes 138 Computer
Coding 139

Index 140
Picture Credits 144

NUMBERS
& SHAPES

The Concept of Numbers

What are numbers? They are a way of expressing very clearly the way we see the world – in essence, they're a bit like words but without the ambiguity. We use them at the simplest level to say things such as, "There are three of us still hungry but only two pieces of pizza left." This one statement tells us three things: the number of people involved in the negotiation (3); the number of pieces of pizza left (2); and the fact that the remaining slices will have to be shared unless someone forks out and buys some more.

Counting Sticks

The earliest example of this basic use of numbers predates watering holes by several millennia. Approximately 10,000 years ago, roughly at the dawn of the Neolithic Age, the tally stick first put in an appearance.

Tally sticks made in England in the thirteenth century, probably salvaged from the fire which destroyed the Palace of Westminster in 1834. These represent examples of the earliest form of bookkeeping.

The earliest known example of a tally stick was a wolf bone. In order to make a permanent record, someone would carve simple, vertical marks into the bone in groups of five. (Why five? Work it out using your fingers.)

This might not sound like much, but at a time when life was short and brutal, it was a major achievement. Unfortunately, that was it for the next 6,000 years or so.

Babylonians and Base 60

The expansion of trade powered the development of mathematics starting about 2000 B.C., particularly in the countries of Babylonia, Egypt, India, and China.

Babylonia, which lay in what is now southern Iraq, was at the crossroads of most of this trade, so it was inevitable that this vital region would see the most rapid development of ideas. By about 1000 B.C. arithmetic had been developed, along with basic algebra and even a primitive form of geometry.

This doesn't mean, of course, that they always got it right. The Babylonians used an unreliable system of counting, called base 60, that is of very little use today (although we continue to divide the minute and hour into 60 units). It was, in fact, the Egyptians who developed base 10, or denary, which we still use today for most of the arithmetic we perform in our daily lives.

Denary System

Base 10, the denary system, uses 10 symbols (0, 1, 2, 3, 4, 5, 6, 7, 8, and 9) to represent all possible combinations of numbers. If, when counting upwards from 1 we reach 9 and then wish to go one further, we simply put a zero in that column (to show that we've moved on) before starting again with a 1 in the next column. This gives us the two-digit number 10.

Binary

Binary, or base 2, works just like the denary system but with just two digits, 0 and 1. Although no one could possible have known it at the time, this was to prove very useful when writing instructions for computers, which are essentially big boxes of switches asking yes/no questions at a remarkably high speed. Binary code, which uses a 1 for "yes" and a 0 for "no," developed out of base 2. See page 139 for more on binary code used in computers.

Infinity

Some numbers are just too big for any of us to comprehend. Even mathematicians eventually get bored of counting, and where they are certain that a number is going to carry on forever they use this symbol, ∞, an eight on its side. See π on page 15 for an example of a number with an infinite number of decimal places.

Pre–modern Mathematics

Modern mathematics might be thought of as starting with the introduction of base 10 and the use of zero as a place-holder. Although it is hard to imagine arithmetic without these two essential elements, people got by for thousands of years without them. This did restrict things. But for the purposes of trade, the use of an abacus provided a way around these problems, not least because by its very nature the abacus is designed with an implied positional system (equivalent sometimes to base 10) and the use of zero as a number.

Developments in Number Systems Through History

When	Where	Development
c. 8000 B.C. (Neolithic Age)	Central Europe (Czech Republic)	First use of a tally stick to record a quantity
c. 2400 B.C.	Sumeria	Place-value system developed
c. 1750 B.C.	Babylonia (Southern Iraq)	First appearance of cuneiform writing to record numbers
c. 1650 B.C.	Egypt	Hieroglyphics first used to notate numbers
c. 1550 B.C.	China	Decimal numbers first used c. 1500 B.C. Bamboo "rods" used to express numbers
900 B.C.	India	First use of zero
300 B.C.	Greece	Euclid writes *Elements*, a 13-volume work on geometry that remains the standard textbook for the next 2,000 years
100 B.C.	China	First use of negative numbers
800 A.D.	Arabia	Birth of algebra

Zero

The first use of zero in the modern sense can be traced back to Muhammad ibn Musa al-Khwarizmi (c. 780-850 A.D.), who is credited as the father of algebra. While the Chinese did not have a symbol for zero, their use of the abacus suggests that they had at least some concept of it. Our own use of zero is tied in with the adoption of the Hindu–Arabic number system, with its numerals (1, 2, 3, 4, 5, 6, 7, 8, and 9) and the notion of zero as both place-holder and "number" in its own right. The Hindu–Arabic system reached Europe via the trade routes in the first millennium and has been the dominant system ever since.

Mathematicians use the word "infinitesimal" to describe a quantity that is less than finite but not quite at zero. Although no such quantity can exist in the real number system, early attempts at developing calculus (see page 55) relied on the use of infinitesimals.

When	Where	Development
1000	Europe	Decimal number system arrives via Arab traders
1514	Holland	First use, in the modern sense, of the symbols + and –
1614	Scotland	John Napier introduces logarithms
1630s	France	Development of analytical geometry
1660s	England	John Graunt lays the foundations of statistics
1660–1670s	England and Germany	Isaac Newton and Gottfried Leibniz develop calculus independently
1830s	Germany	Non-Euclidean geometry developed
1960	U.S.A.	Benoit Mandelbrot develops fractal geometry
1980s	U.S.A.	Chaos theory applied to analysis of complex systems such as the weather

Types of Numbers

Natural Numbers

Natural numbers are the ones we tend to use all the time, particularly when counting things. So, rather than saying "I have X number of gray suits," we say "I have 1 … 2 … 3 … 4 … 5 gray suits." Some mathematicians include zero among the natural numbers, but others do not. This is a cause for continued controversy and is argued widely by mathematicians.

Integers

Integers are natural whole numbers such as 1, 2, 3, etc., together with their negatives, such as −1, −2, −3, etc., and zero. Any addition, subtraction, or multiplication of two integers will always produce another integer. However, this isn't necessarily the case when dividing two integers. For example, dividing the integer 142 by the integer 5 results in 28.4.

Rational Numbers

A rational number is any number that can be obtained by dividing one quantity by another. These include all whole numbers and most fractions.

Irrational Numbers

Irrational numbers are numbers that cannot be expressed as a simple fraction or as a decimal with a finite number of decimal places. The most famous of these is π or "pi" (see page 15).

Figurate Numbers

Some numbers, when expressed as an arrangement of dots, can be formed into geometrical patterns. The ancient Greeks and Chinese were fascinated by this effect, and explored numerous ways of arranging triangular and square numbers, right up to octagonal numbers and nonagonal numbers.

Triangular Numbers

Obvious examples of numbers that are begging to be turned into dots and arranged as (equilateral) triangles include the numbers 3, 6, 10, 15, and so on, which have come to be known as the triangular numbers.

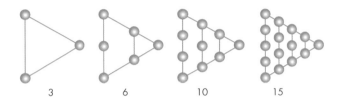

| 3 | 6 | 10 | 15 |

Square Numbers

A square number is any number, such as 4, that can be arranged into a square array of dots.

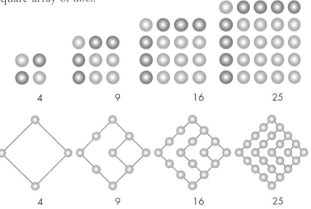

| 4 | 9 | 16 | 25 |

| 4 | 9 | 16 | 25 |

Square Roots

Square roots aren't normally associated with patterns of dots, but they do have a symbol of their own: $\sqrt{\ }$.

If you see a number with a small number two written by it, like this: 2^2, then you know that the number is to be squared, which is to say it is to be multiplied by itself.

For example, 3^2 means that we should multiply 3 by 3, which gives us 9. Having multiplied 3 by itself to produce 9, it follows quite naturally that the square root of 9 must be 3. This might all be written as:

$$3^2 = 3 \times 3 = 9$$
$$\therefore \ \sqrt{9} = 3$$

Factors and Prime Numbers

A number that divides another number evenly, i.e. with nothing left over, is said to be a factor of that number.

For example, the numbers 2 and 5 are factors of the number 10, because 2 divides into 10 exactly 5 times and 5 divides into 10 exactly 2 times.

Eratosthenes's Sieve

The mathematician and astronomer Eratosthenes was clever even by the awesome standards of Greece in the third century B.C. He created a theoretical "sieve" with which to shake out all the numbers that aren't primes in order to find the ones that are.

One quite simply takes out the number 1 before removing every second number after 2 (because 4, 6, 8, etc. can be divided by 2). One then removes every third number after 3, every fifth number after 5, every seventh number after 7, etc. and onwards until one reaches 100. The numbers that remain in the sieve are said to be prime. Eratosthenes, mathematical genius that he was, went on to be the first to attempt, and succeed, to calculate the circumference of the Earth.

However, a number that can be divided only by itself and 1, i.e. a number that has only two factors, is said to be a prime number. Examples of prime numbers include 2, 3, 5, 7, 11, 13, and 17.

Imaginary Numbers

Imaginary numbers are any numbers of the form *ai*, where *a* is a real number other than zero and *i* is the square root of −1.

Complex Numbers

Imaginary numbers combine with natural numbers to produce complex numbers.

π

This symbol, which is pronounced "pie," and sometimes written as "pi" – a letter in the Greek alphabet – is used to denote the ratio of the circumference of a circle to its diameter. It's especially useful in geometry, where it can be used, for example, to help calculate the circumference of a circle. If we know what the diameter of the circle is, we merely multiply this by π to get the circumference of that circle. This gives us the formula c = πd. Similarly, we can find the area of a circle by multiplying the square of its radius by π. This gives us the formula $a = \pi r^2$.

π is an irrational number (see page 12). An approximate figure, of 3, was used in calculations until the third century B.C., when another phenomenal Greek mathematician, in this case Archimedes (of "Eureka!" fame), established a figure of 3.14 for π. This was refined to 3.1416 during the second century A.D. and gradually moved to a greater number of decimal places as the centuries passed.

FACT
In the twentieth century π was calculated to approximately a billion decimal places.

Fractions

Some people's idea of heaven on earth is a Long Island Iced Tea. This cocktail, leaving aside the odd regional variation on the recipe, is made up of roughly equal parts of rum, gin, vodka, tequila, and cola. These five ingredients make up one-fifth each of the whole cocktail. This is a fraction, and it is written down like this: ⅕

This fraction tells us two vital bits of information. The number at the bottom of the fraction tells us how many parts make up the cocktail. In this case, the cocktail (the whole) has been divided into five parts. The number at the top of the fraction tells us how many parts of the whole we have. If we have a fifth of the cocktail then we have just the one shot (and we can only hope and pray that it's not the cola).

⅕ cola

⅕ tequila

⅕ vodka

⅕ gin

⅕ rum

Decimals

The decimal system is widely used for counting and money systems. It uses a combination of 10 different symbols (0, 1, 2, 3, 4, 5, 6, 7, 8, and 9), the value of each of which is dependent on its position. Decimal numbers are grouped in ones, tens, hundreds, thousands and so on. If we assume that our cocktail can be represented by the figure 1 (for the whole) then each individual shot will make up 0.2 of the whole because $5 \times 0.2 = 1$.

Percentages

In mathematics, the word "percentage" is expressed using this symbol, %, which allows us to write down numbers as fractions of 100. If our

FACT

In Roman times order in the ranks was occasionally maintained by the practice of decimation. If there had been a mutiny among the soldiers, punishment was meted out by lining up the entire unit and then, from a random starting point, removing every tenth man from the line. The unfortunate men would then be executed as an example to the others.

whole glass of Long Island Iced Tea represents 100% of the whole, then we can find the percentage of, for example, tequila in the glass simply by dividing the whole by five (there are five equal-sized shots in the glass). This gives us a figure of 20%.

Indices

Indices (index, singular) are a way of indicating that there is more to a number than might first meet the eye. They give us the "power" of a number, which tells us that the true number is actually a multiple of the one we are looking at. In the figure 3^4, the tiny number 4 is the index of 3 and this tells us that the number we are really looking at is $3 \times 3 \times 3 \times 3 = 81$. See page 120 for useful squares and cubes of numbers.

Logarithms

Logarithms ("logs") are similar in concept to indices, but are used to express numbers as powers of 10. For example, the number 100 is equal to 10^2, which means that in this case the log of 100 is 2. For ease of use, tables of logarithms (log tables) have been drawn up, which make it easier to multiply or divide difficult numbers. See page 127 for a useful log table for sine, cosine, and tangent.

Powers of 10

Powers of 10 are a useful way of writing down very large numbers without having to resort to row upon row of zeros. For example, 10 to the first power (10^1) is 10; 10 to the second power (10^2) is 10×10 = 100; 10 to the third power (10^3) is $10 \times 10 \times 10 = 1,000$; and so on. Very small numbers can also be expressed by simply adding a minus sign to the power. For example, 10^{-2} is 0.01.

FACT
The mass of the Earth, when expressed as a power of 10, is 5.978 kg $\times 10^{24}$.
(That's 5,978,000,000,000,000,000,000,000 kg.)

Scientific Notation

Scientific notation depends a great deal on powers of 10 in order to express some of the very large and very small numbers that tend to be used. In truth, it would be almost impossible to work through a complex scientific equation without the use of powers of 10. Even doing something as simple as adding the masses of the rocky inner planets (Mercury, Venus, Earth, and Mars) would prove incredibly long-winded without this useful abbreviation. Just imagine having to write out all of the zeros in this:

3.3 kg $\times 10^{23}$ + 4.87 $\times 10^{24}$ + 5.978 kg $\times 10^{24}$ + 6.4 $\times 10^{23}$ kg
(Mercury)   (Venus)   (Earth)   (Mars)

Dimension

In its simplest terms, dimension can be defined as magnitude measured in a specific direction. This means that a line could be said to have one dimension (length). A surface such as this page has the twin dimensions of length and breadth, while a solid object such as this book has the three dimensions of length, breadth, and height.

Classifications of Two-dimensional Shapes

Triangle

The origins of the triangle lie not in mathematics but in art. Examples of decorative triangles can be found in Sumerian pottery dating from around 3500 B.C. In addition to its mystical and astrological significances, the triangle came to be used in making simple measurements before its other properties were noticed and it became the cornerstone of geometry.

Four-sided Shapes

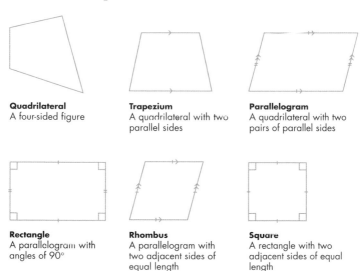

Quadrilateral
A four-sided figure

Trapezium
A quadrilateral with two parallel sides

Parallelogram
A quadrilateral with two pairs of parallel sides

Rectangle
A parallelogram with angles of 90°

Rhombus
A parallelogram with two adjacent sides of equal length

Square
A rectangle with two adjacent sides of equal length

Polygons

Polygons are exotic creatures that inhabit a two-dimensional world. Triangles, squares, and other similar shapes are all examples of polygons. The sides of a regular polygon are of equal length and the interior angles are always the same size. Triangles and squares are classic examples of **regular** polygons. The more sides a regular polygon has, the more closely it resembles a circle. Polygons can be divided into two broad groups: the **convex** and the **re-entrant**. Convex polygons have all of their corners pointing outwards. Re-entrant polygons have one or more corners pointing inwards.

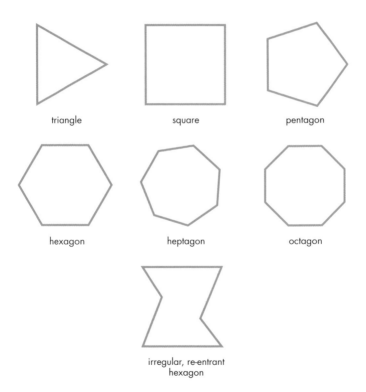

triangle

square

pentagon

hexagon

heptagon

octagon

irregular, re-entrant
hexagon

Curves

One method of creating a curve is to trace the path of a point as it moves. For example, one can create a circle by tracing the path of a moving point that is always the same distance from a fixed point.

Hypatia (c. 400 A.D.) was a Greek mathematician who is famous for her work in the field of geometric curves. Building on the work of Apollonius, she was able to demonstrate that all of the common curves can be created by slicing through a cone. Using this method, which is known as conic sectioning, she was able to form the **circle**, **ellipse**, **parabola**, and **hyperbola**.

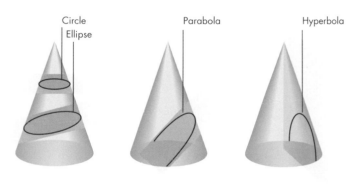

Circle
Ellipse   Parabola   Hyperbola

FACT
Hypatia is almost as famous for the manner of her death as she is for her work on conic sections. Having earlier offended religious leaders, she was pulled one day from her chariot by a gang of monks and dragged into a church where she was tortured to death before being quartered and burned. (Some say that mathematicians have been getting their revenge on the rest of humanity ever since.)

Circle

The outer edge of a circle is called the **circumference**. The distance from the center of a circle to its circumference is called the **radius**. This is equal to half the **diameter** of the circle.

Dividing the circumference of a circle by its diameter results in an irrational number π (see page 15 for more on π). This figure is exactly the same for every circle and is approximately equal to 3.142.

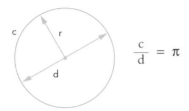

$$\frac{c}{d} = \pi$$

Solid Figures: Polyhedrons

The properties of solid figures have kept mathematicians occupied for centuries. Sometimes called polyhedrons, they are formed from regular polygons, such as squares and triangles, and, despite their best efforts, mathematicians have so far failed to find any more than five of them.

These five solid figures have been known since ancient Greek times, which is why they are sometimes referred to as the **Platonic solids**. The five regular polyhedra are the **tetrahedron** (four triangular faces), the **cube** (six square faces), the **octahedron** (eight triangular faces), the **dodecahedron** (12 pentagonal faces), and the **icosahedron** (20 triangular faces).

Sphere

The sphere can be thought of as a circle in three dimensions. Every point on the surface of a sphere lies at the same distance from its center. As with a circle, this distance is known as the radius. Also just like a circle, a sphere has a circumference and a diameter. Slicing through the diameter of a sphere creates two equal **hemispheres**, the flat surfaces of which form circles.

Following are a couple of useful formulae associated with the sphere that the Greek mathematician Archimedes figured out over 2,000 years ago:

$$\text{Surface area of a sphere} = 4\pi^2$$
$$\text{Volume of a sphere} = \tfrac{1}{3}\,\pi r$$

See page 121 for more on the surface area and volume of three-dimensional shapes.

Pyramids

The classic Egyptian-style pyramid has triangular sides, but unlike the tetrahedron it has a square base. Slicing an Egyptian pyramid parallel to its base produces a square face, as demonstrated in the diagram below.

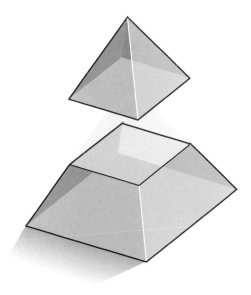

Topology

Topology is a fun area of mathematics and unlike many others is really quite a recent development. Very loosely, topology is the name given to the study of the properties that an object retains when it is **deformed**. (Deformed in this sense means either bent, stretched, or squeezed.) An easy way to visualize this is to imagine a circle being pushed and shoved until it takes on the shape of a triangle. These two shapes, the circle and the triangle, are said to be topologically equivalent.

Königsberg Bridge Problem

The origins of topology lie in attempts to solve the Königsberg Bridge problem. Königsberg, a town which in the early 1700s lay in Germany, is built on a river. There is a small island in the middle of the river which is connected via seven bridges to the rest of the city. The challenge, which clearly pre-dates the invention of the television, was to find a way to cross each of the seven bridges in a single journey, returning to the original starting point without retracing any steps.

Despite the best efforts of all concerned, the solution to the problem was never found as it is impossible to cross all the bridges in a single journey without retracing one's steps.

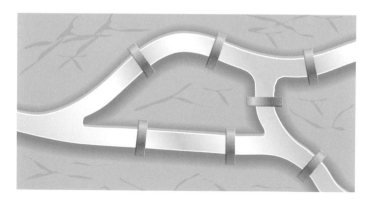

Make Your Own Möbius Strip

The Möbius strip looks like something out of a drawing by Escher but can be made in the real world. In fact, the Möbius strip has been used to provide drive belts for power-driven machinery for quite some time. You can make your own Möbius strip by fixing two ends of a rectangular strip of paper together having first given one end a half twist.

Despite initial appearances, the Möbius strip has only one edge and one side. Also, if you split a Möbius strip down the middle, it remains in one piece. These unusual qualities were first discovered independently but simultaneously in 1858 by the German mathematicians August Möbius and Johann Listing – but clearly Möbius received the credit for it.

FACT

In the strange world of topology there is no real difference between a coffee cup and a topologically equivalent donut. One holds coffee, the other jam, but both can have the same topology.

The Beauty of Numbers

The Golden Mean

Some things just look right, be it the arrangement of the furniture in a room or the composition of the various lines and shapes that go to make up a painting. Being essentially two-dimensional, the patterns of lines and shapes that form the images are of vital importance, even for those artists who specialize in conveying a sense of depth. Over time, artists noticed that some arrangements worked better – were more pleasing to the eye – than others.

Classic styles of painting rely heavily on striking a perfect balance between the proportions of the parts to the whole, a problem that, of course, drew the attention of various mathematicians. The Golden Mean (or Section) was defined by the Renaissance mathematician Lucas Pacioli, who claimed that it was the "divine proportion." It refers to the ratio of the division of a line so that the shorter part is to the long as the long is to the whole. The ratio is approximately 8 to 13. He published his ideas in a treatise called *Divina proportione*, which was said to have influenced Leonardo da Vinci.

Spirals

Like triangles, the origin of man-made spirals lies in the world of art, where they were used, particularly by the Celts, to provide decoration. The first mathematician to formalize our understanding of the spiral was Archimedes, who has a spiral named for him. The formula for an **Archimedes's spiral** is $r = a\theta$, where *r* is the length of the radius, *a* is a constant and θ is the amount of rotation (or "angular position," as Archimedes called it).

Another type of spiral is the **equiangular**, or **logarithmic**, **spiral**, which was discovered in the seventeenth century by the French mathematician and philosopher René Descartes. This particular spiral is surprisingly common in nature, and can be seen in spiders' webs and, to a greater degree of accuracy, in the Nautilus shell of the chambered mollusk. Spirals can also be found in certain flowers. The giant sunflower has florets arranged in two intersecting spirals, the clockwise arrangement having 34 spirals and the counterclockwise having 55 spirals. Interestingly, both of these numbers (34 and 55) are part of the Fibonacci Series (see page 28).

Undoubtedly, the largest spirals in nature occur out in space, where truly massive spiral galaxies can be observed in all their fiery majesty.

Spiral galaxy NGC 4414 as imaged by the Hubble Space Telescope in 1995. This galaxy is 60 million light years away from our own.

FACT

Own own home galaxy, the Milky Way, is a truly gigantic spiral arranged around a massive black hole. The spiral is about 100,000 light years across, up to 5,000 light years thick and rotates at a rate of around once every quarter of a million years.

Fibonacci Series

An interesting sequence of numbers, which has parallels in nature, was first spotted in 1202 A.D. by the medieval Italian mathematician Leonardo of Pisa (known also as Fibonacci). He attempted to answer the question "How many pairs of rabbits can be produced in one year if each pair produces a new pair *which become productive from the second month*?" Leonardo knew that one pair of rabbits would produce another pair (1, 1) and that by the end of the second month the two pairs of rabbits would produce an additional two pairs between them (1, 1, 2), and so on.

This eventually produced a number sequence where each new addition is the sum of the previous two numbers, giving us the pattern 1, 1, 2, 3, 5, 8, 13, 21, 34, 55, etc. This is now known as the Fibonacci series.

Surprisingly, the Fibonacci series occurs quite naturally in nature, often tallying with the numbers of petals on flowers or segments in a pinecone (illustrated below). Even the pineapple is a living monument to Fibonacci, with eight rows of scales sloping to the left and 13 to the right.

USING
NUMBERS

The Development of Measuring Systems

For obvious reasons, body parts provided the first reference points for standard measurements. The cubit, for example, was based on the distance from the base of the elbow to the tip of the extended fingers. It was used throughout the Middle East in ancient times.

Similarly, early weights were based on the amount a person, or an animal, could carry. Over time, the practice of using roughly standardized weights and measures drifted westward as trade carried standard weights and measures to Greece and then on to the Roman Empire. Although they developed independently of the western system, Chinese measuring systems also resemble closely those of the Mediterranean in this period.

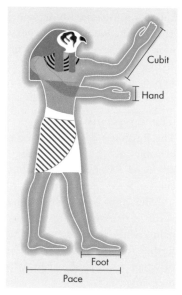

The Egyptian cubit became the first standardized measure in about 3000 B.C.

Babylonian Measures

The earliest-known standard weight was the Babylonian **mina**. Babylonia was also where the **shekel**, often thought of as a Hebrew coin, originated as a unit of weight. The Babylonians had their own cubit which, at about 530 millimeters, was slightly longer than the Egyptian one. They also had a liquid measure, called the **ka**, which was equivalent to the volume of a cube with sides of 100 millimeters.

The measuring systems developed by the Babylonians and the Egyptians were generally adopted and adapted by the Hittites, Assyrians, Phoenicians, and the Hebrews. True standardization didn't occur, however, until the first century B.C., when the Greeks came to dominate trade in the Mediterranean. They eventually ceded power to the Romans who adapted the Egyptian system which, through a combination of conquest and ubiquity, spread throughout their empires and beyond.

Roman troops were renowned for their ability to march 20 miles a day in full battle armor, so it is surely no accident that the true legacy of the Roman system lies with the measurement known as the **foot**. Based, obviously, on the length of a human foot, this was further divided into 12 **inches** (called *unciae*). Five feet was equal to a **double step**, or **pace**. A thousand paces made up a single Roman **mile** (*mille passus*), the first time such a distance was formally standardized.

The Egyptians

The Egyptians developed their cubit in about 3000 B.C. This later became the first standardized measure when Egyptian royalty produced a black granite master cubit against which all Egyptian cubit sticks were measured.

The cubit was subdivided into 28 **digits**, which were probably equal to the width of one finger. Four digits were equivalent to a **palm**, while five made a **hand** and 14 made a **span**. The digit was further subdivided to produce fraction-like measures, the smallest of which was $\frac{1}{16}$ of a digit, which was in turn equal to $\frac{1}{448}$ of a royal cubit.

Although these may not sound like the most reliable measures, the Egyptians used them to great effect – and with remarkable accuracy – when building their pyramids.

FACT
Despite its huge size and the thousands of people involved in building it, the sides of the Great Pyramid at Giza in Egypt are accurate to within 0.05 percent of each other.

The Middle Ages: The Pound, Foot, and Stone

The days of yore were followed by the Dark Ages, which eventually gave way to the Middle Ages. This period saw the Roman system, complete with its Egyptian, Babylonian, and Greek influences, firmly established in Europe, although by this time elements of the Arabic and Scandinavian systems had also exerted an influence. For example, the basic Roman unit of weight, the **libra**, became known as the pound, but retained its abbreviation, *lb*, thereby betraying its Roman origins.

Trade fairs in the twelfth and thirteenth centuries, which saw merchants from all over Europe coming together to do business, had a significant impact on the move towards standardization. In Britain, the Magna Carta of 1215 saw the first steps towards standardized measures that would remain in place for nearly 600 years. This period saw the **yard** established at 3 feet, each of which was further divided into 12 inches, although the seeds of confusion were sown when the English **stone** was defined as having 14 pounds (no one likes working in units of 14).

In a so far half-hearted — and really rather late in the day — attempt to improve things, the British rushed through an act of Parliament in 1963 under which the population was supposed to adopt the metric system. Nearly 800 years after the Magna Carta, there is still a considerable degree of resistance across Britain to any attempt to make things clearer and simpler.

FACT

In the Middle Ages the taxman adopted the use of an abacus-like chequered cloth on which to move counters in order to keep track of who owed what. The use of this chequered cloth is still reflected in the title of the person who has the ultimate responsibility for collecting taxes in Britain, the Chancellor of the Exchequer.

Revolution in France: The Metric System is Born

The French Revolution of 1789 had, and continues to have, a significant impact on the rest of the world. In the midst of the revolution, the French introduced a new and entirely rational system of weights and measures. In 1791 a committee of scientists got together and began work on establishing what would become the metric system of measurement.

In 1793 the standard **meter** was defined and in 1799 the metric system, with its meters, **grams**, and **liters**, was formally adopted throughout France. Founded on the very logical base 10 system, metric measurement spread across Europe in a pattern broadly equal to that carved by the invading French armies under Napoleon. Even after Napoleon's death, the system continued to spread, being adopted by the Japanese in 1868. The United States signed the Metric Convention of 1875 and adopted the metric system for scientific purposes, but, like the British, retains use of the old Imperial system in everyday life, with its pounds, ounces, and accompanying confusion.

SI Units

Because science and scientists made such enormous strides forward during the period in which the metric system was introduced, it soon became apparent that the new system was not quite up to the job of coping with the demands placed on it by modern science.

In October 1960, the eleventh General Conference on Weights and Measures met in Paris to draw up a revised metric system, which has been altered and amended from time to time ever since. This new International System of Units (or SI, as it is usually called) redefined standard metric measurements to a higher degree of accuracy by using developments in science, such as the establishment and recognition of the constant speed of light in a vacuum. For the most part, artifacts such as a standard metric rule were abandoned in favor of this far more accurate and consistent measure. See pages 124–126 for tables of SI measurements and definitions.

Time Measurement

Self-awareness is what distinguishes us from the rest of the animal kingdom. We are aware that we have been born, can expect to live for a while, and will eventually die. We are able to surmise this because we are aware of the passage of time. Time has a profound effect on our lives and so it was inevitable that we would attempt to find ways of measuring it, if only to discover how much longer we have left.

Even in ancient times, few people could have failed to notice the rhythmic passage of the seasons, the phases of the Moon and the journey across the sky of the wandering stars (as the planets were originally known). Some of the earliest timepieces relied on the apparent movement of celestial bodies such as the Sun in order to mark the passage of time.

Sundials

The earliest forms of the sundial date from around 3500 B.C. They used a stick, called a **gnomon**, to cast a shadow across a dial which had been calibrated to tie in with the movements of the Sun. Useful on sunny days only, the sundial proved to be little more than an interesting ornament when the Sun didn't shine or night-time fell.

Candles

Candles that were known to burn at a steady rate were adapted to display the passage of time. Bands were marked down the side of the candles at roughly one-hour intervals.

Phases of the Moon: New Moon, Waxing Crescent, First Quarter, Waxing Gibbous

Pendulum Clocks

The pendulum marks the point at which measurements of the passage of time began to become a little more accurate. A pendulum always takes the same amount of time to complete a cycle, almost regardless of the speed or duration of the swing. As the pendulum slows so the length of the swing shortens, which compensates for the drop in speed. This means that a release mechanism tied to the swing of the pendulum will operate … well, like clockwork.

Quartz Digital Clocks

Passing an alternating current through a piece of quartz crystal causes it to vibrate millions of times every second. This vibration is consistent and can be used as a base against which to measure the passage of time. Quartz technology developed in the 1930s and 40s and improved timekeeping performance far beyond the the pendulum.

Atomic Clocks

Atomic clocks rely on the resonant frequency of certain atoms to help in the measurement of the passage of time with extreme accuracy. The energy change of an atom produces a regular pulse, which can be measured, quantified, and counted.

The current SI standard for the passage of one second is the time taken for a cesium-133 atom to cycle through 9,192,631,770 of these energy changes.

Full Moon, Waning Gibbous, Last Quarter, Waning Cresent

Greenwich Mean Time and Universal Time

Greenwich Mean Time (GMT) was calculated from 0° longitude at the Royal Observatory at Greenwich in England. The plan was to avoid confusion over variations in local time by calculating it from this one spot. Originally, 00.00 GMT was at noon, but this was changed in 1925 so that the day began at midnight. Unfortunately,

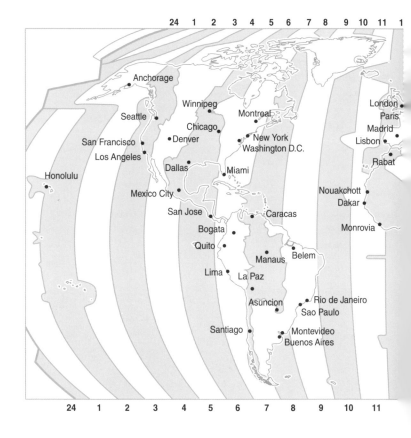

not everyone kept up and confusion reigned once again. In 1928 the International Astronomical Union decided to replace Greenwich Mean Time with Universal Time, which for practical purposes differs from GMT in name only. The term "GMT" is still used by English-speaking navigators.

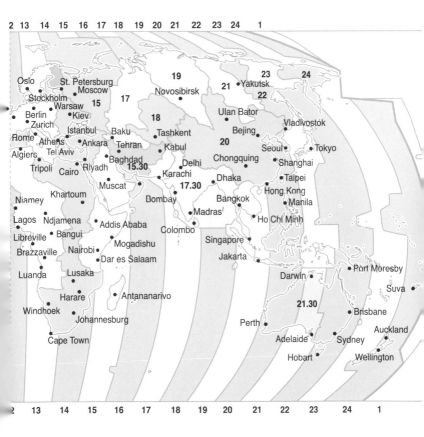

Astronomical Measurements

The Universe is incomprehensibly vast, a fact that was not truly appreciated until relatively recently. In order to interpret the vastness of space mathematically, it became necessary to figure out new ways of measuring it. Because we have a good idea of the speed of light in a vacuum (and most of space appears to be a vacuum), it seemed that the distance traveled by light during a set period of time might be a useful way of measuring distances in space.

Spacial units	Definition
Speed of light	186,212 miles (299,792 km) per second
Light year	The distance traveled by light in a year = 5.878×1012 miles (9.46053×1012 km)
Astronomical Unit (AU)	The average distance from the Earth to the Sun = 92,955,808 miles (149,597,870 km)
Parsec	The distance at which the radius of the Earth's orbit subtends an angle of one second of arc = 3.26 light years
1 kiloparsec	1,000 parsecs
1 megaparsec	1,000,000 parsecs

Planets in our Solar System from closest to furthest from the Sun: Mercury, Venus, Earth, Mars, Jupiter, Saturn, Uranus, Neptune, Pluto.

FACT

If you laid out a scale model of our Solar System so that it fitted onto the Center Court at Wimbledon, with the Sun at one end and Pluto at the other, the nearest star to our Sun (Proxima Centuri) would have to be placed in Johannesburg, South Africa, in order to keep the model to scale.

Theory of Measurement

Since the beginning of time humankind has found a need for accurate measurement in our daily lives. The first person to develop a theory of measurement was the Greek mathematician Eudoxus, whose work was included in Euclid's *Elements*.

Things have moved on a little since then and nowadays scientific measuring systems are calibrated across a vast range, covering everything from the number of protons in the nucleus of an atom to the brightness of stars in the night sky.

Scientific Measurements

Atomic Number

The atomic number of an atom is measured by determining the number of protons it contains in its nucleus. Hydrogen, for example, has just one proton so its atomic number is one.

Atomic Mass

The atomic mass of an atom is measured by determining the numbers of protons and neutrons in its nucleus. Carbon has six protons and six neutrons, which gives it an atomic mass of 12.

Molecular Mass

Also known as the **relative atomic mass**, this figure represents the mass of a molecule calculated relative to one-twelfth the mass of an atom of carbon-12. It is found by adding the relative atomic masses of all of the atoms that make up the molecule. Most molecules contain fairly small numbers of atoms, but some have rather more. Certain molecules of rubber may have up to 65,000 atoms.

Apparent Magnitude

This is a measure of the brightness of a star as it appears from Earth. Extremely bright stars are said to have a magnitude of one, while very faint stars are said to have a magnitude of six.

Absolute Magnitude

This is a measure of the magnitude of brightness of any given star as if viewed from a set distance of 10 parsecs (32.6 light years).

Decibel Scale

The decibel (dB) scale is used to measure the intensity of sound.

Decibels	Sound level
0	The limit of audible sound
10	Low whisper
20	Normal whisper
20–50	Quiet conversation
50	Normal speech
50–65	Loud conversation
65–70	Traffic on busy street
70–90	Passing train
75–80	Factory
90–100	Thunder
110–140	Jet plane taking off
140–190	Space rocket lifting off

Range of the Electromagnetic Spectrum

Type of radiation	Frequency
Radio	Up to 3,000 MHz
Microwave	3,000 MHz to 3,000 GHz
Infrared	3,000 GHz to 430 THz
Visible light	430 THz to 750 THz
Ultraviolet	750 THz to 300 PHz
X-rays	300 PHz to 30 EHz
Gamma rays	More than 30 Ehz

Simple Graphs and Charts

Pie Chart

A pie chart is one of the simplest ways of presenting data so that it can be easily absorbed. It relies on dividing a pie-like graphic into segments, the size of each of which is a reflection of the value it represents. Pie charts are especially useful for displaying limited amounts of information, such as the answer to the question "People from which age groups are watching a particular TV network at the prime time of 9 p.m.?" (when the advertisers come out to play). The combined values of all of the segments in a pie chart should add up to 100 percent.

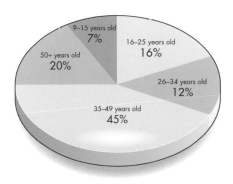

Line Graph

A line graph is slightly more complex than a pie chart because, in effect, it has an extra axis. This allows for the addition of other information, much of which tends to be time-related. Once all of the information has been plotted, such as in this example, which shows the number of items sold in different months of the year, the data points can be joined by a broken line or clustered around a line or curve.

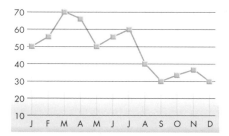

Scatter Diagram

A scatter diagram is essentially a line graph without the line but with all of the individual bits of information plotted as points on a graph. There is then an often irresistible temptation to draw a line through the area of the graph where there is the greatest concentration of points. Doing this turns a scatter diagram into a line graph.

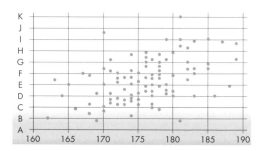

Bar Chart

A bar chart represents information in strips or "bars" which vary in height according to the value they represent. Similar to a line graph, it plots information in a way that often gives a Gaussian curve (see page 45).

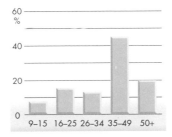

3D Graphs

Improvements in technology, particularly those associated with computing and computer graphics, mean that it is easier than ever to generate three-dimensional graphs and charts. One example of use of this type of graph is in plotting topology. Another striking example of this involves the application of computer graphics to a sound wave monitor. This allows the user to take a "snapshot" of a particular moment in time and display all of the properties of the sound wave. Continuous monitoring can also be carried out, allowing the user continually to see the size, shape, and intensity of the sound wave.

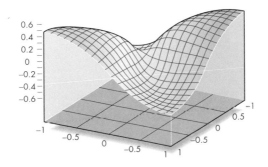

Probability and Statistics

Probability is the branch of mathematics that is concerned with chance. What is the chance of the Red Sox winning the World Series? What chance does Ted Kennedy have of becoming the next U.S. president? These are questions that cause us to examine the probability of an event occurring. The answers, in each case, will lie somewhere between one (a racing certainty) and zero (no chance) and are expressed most often as a fraction or percentage.

Probability, as a mathematical tool, is used in a branch of mathematics called statistics.

Statistics begins with the collection of data, which is then analyzed to identify trends and predict likely outcomes. Often, the findings from this analysis are displayed using graphs and charts (see pages 41–43), which present the information in a way that is easy to absorb. For example, by importing data into a line graph it might be possible to see where the average in that set of data lies. This average is called the mean, but perhaps confusingly it is not the only kind of average statisticians use. In fact, they have three types of average: the **mean**; the **mode**; and the **median**.

Mean

The mean is found by dividing the sum of quantities by the number of quantities. For example, if you divide the number of glasses of wine drunk at dinner by the number of guests present you will find the mean. This is, if you'll excuse the pun, what most people mean when they say "average."

Mode

In almost any set of figures, the number which occurs most often is called the mode. So, if you have six dinner guests and three of them admit to drinking three glasses of wine each with dinner, while two guests had half a glass each and the other drank two, then the mode is three (glasses of wine).

Median

The median is the middle number in a set of numbers that have been arranged in order of size. If there are two middle numbers then the median is the average of the two figures.

Gaussian Curves

Generally speaking, a line graph or bar chart showing the distribution of data around an average value will produce a distinctive-looking curve. This is known as the Gaussian curve and it is produced by random variation around the mean.

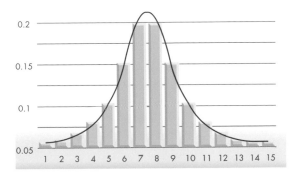

FACT

Statistically speaking, in every war from the end of the fifteenth century until the start of the Second World War, you were more likely to die from typhus than from armed assault by the enemy.

Stats in Socks

There you are, late for work and still not wearing any socks. In front of you lies a bag of clean laundry containing five black socks, three gray socks, and a pair of socks with a cartoon character emblazoned on the ankles. Without emptying the bag, you select your socks at random.

The initial probabilities of selecting either a black sock, gray sock, or cartoon sock are 5/10, 3/10, and 2/10 respectively. As luck would have it, you pull out a black sock on your first go (hurrah!). This means that the probability of you pulling out another black sock is now 4/9. However, you may, of course, pull out a cartoon sock, or perhaps it will be a gray sock. To find the likelihood (probability) of this, you must multiply the two individual events together. Or, of course, you could simply put your hand in the bag and pull out another sock.

Stocks and Scares

Probability is of paramount importance to two groups of people: gamblers and stockbrokers. Essentially, these two groups of people play the same game. A gambler will examine a horse's track record and will, taking into account other factors such as the jockey and the course conditions on the day, decide whether or not to place a bet depending on the probability of the horse winning. A stockbroker does pretty much the same thing; it's just that the bets are bigger and the consequences of mistakes somewhat closer to catastrophic.

ALGEBRA &
TRIGONOMETRY

Algebra

Algebra is one of the branches of mathematics that uses letters and symbols, as well as numbers, to express the various parts of an equation. Algebra is essentially a hunt for the values that lie behind the letters and symbols in equations.

Algebraic expressions normally contain both **constants** (parts that have a fixed value) and **variables**. An example of this is the equation for the circumference of a circle:

$$c = 2\pi r$$

In this equation, the variables are c (the circumference) and r (the radius of the circle); 2π is the constant in this equation.

An equation, by definition, has to have sides of equal value (that is, after all, what the "equals" sign means). Where an algebraic expression has sides of unequal value, the symbol > is used to mean "greater than" while the symbol < is used to express "less than."

FACT

Algebra gets its name from the ninth-century Islamic mathematician Muhammad ibn-Musa al-Khwarizmi, who wrote *The Book of Restoring and Balancing*. (The Arabic word for "restoring" is *al-jabr*.)

A Few Useful Definitions

An algebraic expression comprising a combination of letters and numbers that includes multiplication and/or division but no addition or subtraction is said to be a single term, or **monomial**. (Each combination of letters and/or numbers separated by + or − is called a **term**.)

An algebraic expression that includes a value that is to be added or subtracted is said to consist of two terms and is a **binomial**, such as:

$$2m + 3$$

An algebraic expression with three terms is said to be a **trinomial**, such as:

$$a + b + c$$

An algebraic expression involving more than three terms is called a **polynomial**, such as:

$$a + b + c + d$$

Types of Equation

A basic equation consists of two or more parts that are equal, such as:

$$2a - 5 = 27$$

A **quadratic equation** is an equation that contains a squared term, such as:

$$x^2 + 2x - 15 = 0$$

Simultaneous equations involve two or more equations happening at the same time, with the hunt being on to find the numbers that are hiding behind two or more letters, such as:

$$2a - b = 5$$
$$3a + 2b = 18$$
$$\text{(i.e. } a = 4 \text{ and } b = 3.)$$

FACT
The Pythagorean Theorem was almost certainly thought up by one of his followers rather than by Pythagoras himself.

Pythagorean Theorem

The Pythagorean Theorem states that the square of the length of the hypotenuse of a right-angled triangle is equal to the sum of the squares of the other two sides (the adjacent side and the opposite side). This is stated in the following equation:

$$C^2 = A^2 + B^2$$

Armed with this information – and the lengths of two sides of a right-angled triangle – one can calculate quite easily the length of the missing side.

In the diagram below, the square of the length of side C (the hypotenuse) is equal to the combined squares of side A and side B. If the square of side A is 16 cm, and the square of side B is 9 cm, then it follows that the square of the hypotenuse is equal to 16 cm + 9 cm, giving us a figure of 25 cm. From here it is a simple step to find the square root of 25 cm and establish that the length of side C is in fact 5 cm.

$$C^2 = A^2 + B^2$$
$$25 = 16 + 9$$
$$C = 5$$

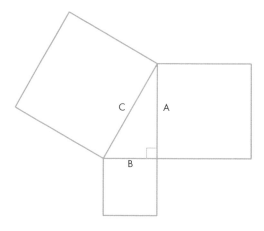

Fermat's Last Theorem

Fermat's last theorem is a theorem first proposed by Fermat in the form of a note scribbled in the margin of his copy of the ancient Greek text *Arithmetica* by Diophantus. The scribbled note was discovered posthumously and the original is now lost. However, a copy was preserved in a book published by Fermat's son.

Fermat's last theorem states the following:

$x^n + y^n = z^n$ has no non-zero integer solutions for x, y, and z when n > 2.

Fermat did not leave any proof and this theorem was not found until 1993.

Coordinate Systems

Coordinate systems are a useful way of plotting the positions of points in space and time. The French philosopher René Descartes gave us one of the most straightforward coordinate systems. It consists of two lines, called axes, placed at right angles to each other. Sets of numbers, called coordinates, are used to define points within the coordinate system.

Each coordinate in the Cartesian coordinate system is made up of two figures. These reflect a position on the **x axis** (the horizontal) and the **y axis** (the vertical). **Vectors**, which are lines reflecting both magnitude (size) and direction, are represented within a coordinate system as lengths and angles.

Vector Coordinates

The movement of a point in a certain direction and over a certain distance is called a vector. When plotted on a map, the vector is the line that might in real life be referred to as the route that the crow flies. In the example over the page, the journey begins at point A and ends at point B. However, there is a stop-over at point C. So, although the actual journey takes us along the path ACB, which is indicated by the dotted line, the vector is actually line AB. The magnitude of vector AB can be found using The Pythagorean Theorem.

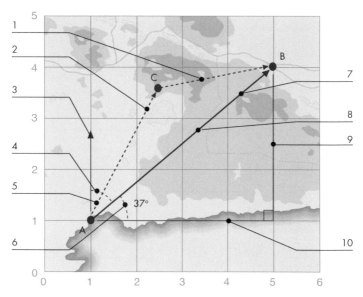

1 Vector CB

2 Vector AC

3 Direction of North

4 Compass bearing is angle clockwise from North

5 Compass bearing is 53° (90°–37°)

6 Angle of vector measured from the horizontal (37°)

7 Magnitude of AB is 5 units, working out using Pythagoras's theorem

8 Length of opposite side is difference between y coordinates of points

A and B, which is equal to 3 units

9 Vector AB found by adding AC+ CB

10 Length of adjacent side is difference between x coordinates of points A and B, which is equal to 4 units

Graphs and Functions

In the world of mathematics, any relationship involving two or more variables is called a **function**. Plotting a function in a Cartesian coordinate system produces a graph, known as the **graph of the function**.

Following are the graphs of the functions for sine, cosine, and tangent, each of which has been plotted in a coordinate system.

Sine

$$\sin \theta = \frac{\text{opposite}}{\text{hypotenuse}}$$

This function can be expressed with the equation y = sin x, which produces the following curve when plotted on a graph.

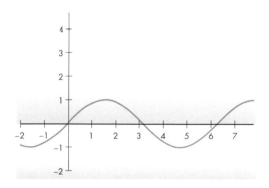

Cosine

$$\cos \theta = \frac{\text{adjacent}}{\text{hypotenuse}}$$

This function can be expressed with the equation y = cos x, which produces the following curve when plotted on a graph.

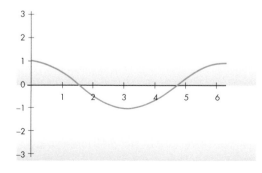

Tangent

$$\tan \theta = \frac{\text{opposite}}{\text{adjacent}}$$

This function can be expressed with the equation y = tan x, which produces the following curve when plotted on a graph.

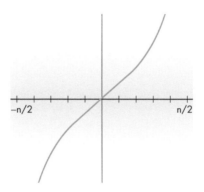

Polar Function

Another example of a coordinate system, which one does not find in everyday math, is the polar coordinate system. Polar coordinates are used to express complex numbers (see page 15) as points in a system. The function is expressed as the equation z = r (cos θ + i sin θ), the graph of which is:

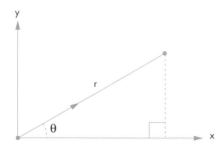

Calculus From First Principles

Calculus! Now there's a word that strikes fear into the very heart of functionally innumerate liberal arts graduates. So how bad can it be?

Calculus is a very useful branch of algebra that emerged in the 1680s. It was developed independently by a German mathematician called Gottfried Leibniz and an English guy. Unfortunately for Leibniz, the English guy happened to be Isaac Newton, arguably the greatest mathematician and scientist in history. Convinced – not entirely unreasonably – that his ideas had been stolen, Newton spent the next two decades hounding Leibniz to his grave.

In truth, Leibniz had been influenced by Newton. Newton's publisher, Collins, had shown the great man's work to Leibniz in the 1670s. But although Newton's approach was superior, Leibniz's mathematical notation had real elegance and lives on to this day.

The Concept of "Limit"

A fundamental principle in the calculus is that of "limit." But this is not a new idea. The ancient Greek mathematicians were familiar with the concept. When Archimedes set out to find the area of a circle he didn't have the useful formula that we use now – he still hadn't figured it out yet. He had, however, found a way to work out the area of a regular polygon.

So he began by drawing equilateral polygons with ever more sides inside the circumference of the circle until he approached the threshold, or "limit," required to define the area of the circle. As the number of sides increased – and so came closer to resembling the circle – so the accuracy of his calculations improved. He eventually arrived at the formula $a = \pi r^2$ in which r is the radius of the circle. The same technique, using rectangles – the areas of which are found by multiplying base by height – can be used to find the areas of irregularly shaped objects. This technique was also applied to finding the volumes of solid objects such as spheres and cones. But the real beauty of the calculus is that it provides a systematic method for arriving at an exact calculation of areas, volumes, and lots of other things without messing about with polygons and rectangles.

Calculus is not, however, concerned simply with fixed quantities such as areas. It is an incredibly powerful tool for calculating continuously varying quantities.

Differential Calculus

Imagine for a moment that the apple of legend really did fall on Newton's head and that he then grabbed it and threw it away so that its path described an arc as it flew through the air (and eventually fell to earth). Using calculus, we can figure out its velocity and its rate of acceleration after it left Newton's hand. This is because, a little like Archimedes and his ever-more-sided polygons, calculus is about infinitely small changes in continuously varying quantities.

Working on the perfectly valid assumption that at any point up to the end of its journey the apple moves a vanishingly small additional distance during an equally small amount of time, we can begin to ascribe values to these two quantities.

If we call the distance traveled dx, the time taken to do this dt, and the velocity of the apple v, this gives us the equation:

$$v = \frac{dx}{dt}$$

We can then manipulate the equation to find the acceleration of the apple.

$$\frac{d^2x}{dt^2} = \frac{dv}{dt} = a$$

The technique for making these calculations is called Differential Calculus, which is one of the basic techniques of calculus.

Integral Calculus

The other basic technique is Integral Calculus, which can be used to find the area enclosed by the curve of the apple's path and its position at any time during its journey across the sky.

In real life, calculus is probably the most widely used branch of mathematics. Being concerned with the relationship (function) between variables, it can be used to calculate solutions to even the most mundane of real-world problems, such as the safe place to put a handle on an oven door (by calculating temperature as a function of the distance from a heat source). But it is its use in computer modeling, for anything from complex engineering construction projects to investigating the ideal acoustic properties of a concert hall, that has seen Newton's work become ever more relevant in the modern world.

Properties and Types of Angles

1 Angles are formed at the point where two straight lines meet.

2 Angles are measured in units called degrees, which are denoted by the symbol °.

3 A single degree represents one 360th of the circumference of a circle. There are 360 degrees in a full circle.

4 The internal angles of a triangle always add up to 180°.

5 A perpendicular bisector is a line that cuts another line exactly in half at right angles.

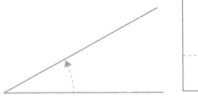

6 An **acute angle** measures less than 90°.   **7** A **right angle** measures exactly 90°.

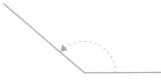

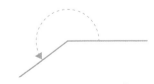

8 An **obtuse angle** measures more than 90° but less than 180°.

9 A **reflex angle** measures more than 180° but less than 360°.

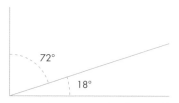

10 Complementary angles always add up to 90°.

11 Supplementary angles always add up to 180°.

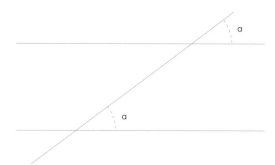

12 Corresponding angles (between parallel lines) are always equal.

FACT

While painting the Sistine Chapel, Michelangelo was approached by the Pope and a conversation developed during which the two men debated the nature of artistic talent. When asked by the Pope for a demonstration of his legendary abilities, Michelangelo took hold of a piece of paper, picked up a pencil and – freehand – drew a perfect circle.

Parts and Properties of a Circle

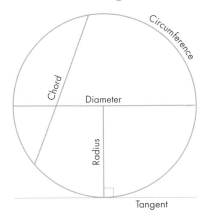

Circumference: The perimeter of a circle

Diameter: A line which bisects a circle, passing through its center

Radius: The distance from the center to the circumference of a circle

Chord: A straight line joining two points on the circumference of a circle

The **tangent** of a circle is perpendicular to its radius

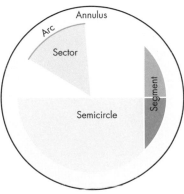

Arc: Part of the circumference of a circle

Sector: Part of u circle bounded by two radii and an arc

Segment: Part of a circle bounded by a chord and an arc

Semicircle: Part of a circle bounded by the diameter and an arc

Annulus: The space between two concentric circles

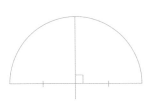

Any line bisecting a **semicircle** always forms a right-angle.

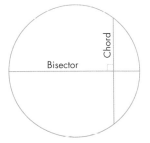

The **perpendicular bisector** of a chord always passes through the center of the circle.

Trigonometry

Trigonometry is concerned with the relationships that exist between the sides and angles of triangles. At a practical level, being able to work out the "missing" length of the side of a triangle (or a "missing" angle within the triangle, for that matter) is enormously useful for all sorts of people, including builders, surveyors, engineers, astronomers, and navigators.

Trigonometry was, like so many things, developed by the ancient Greeks. This is reflected in the name given to this branch of mathematics, which is derived from the Greek *trigonon* (meaning triangle) and *metron* (meaning measure). While it's obvious that the Egyptians knew a thing or two about triangles – just look at those pyramids! – the person credited with laying the foundations of trigonometry is Hipparchus, a Greek astronomer from the second century B.C.

Types of Triangle

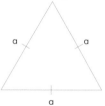

An **equilateral triangle** has sides of equal length and internal angles of equal size (60°).

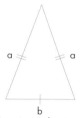

An **isosceles triangle** has two sides of equal length and two internal angles of equal size.

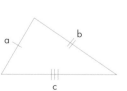

A **scalene triangle** has no sides of equal length and no internal angles of equal size.

A **right-angled triangle** has one internal angle of 90°.

A Few Things You Should Know About Triangles

A triangle is a plane (two-dimensional) figure with three sides. It has internal angles which add up to 180°. In a right-angled triangle, the side opposite the right-angle is called the **hypotenuse**. The side opposite the angle θ is called the **opposite**, while the remaining side, i.e. that which is neither the hypotenuse nor the opposite, is called the **adjacent** side. Each side is associated with an identically named angle.

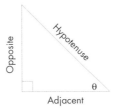

Sine, Cosine, and Tangent

Sine (sin) θ $= \dfrac{O}{H}$

The sine of the angle θ is the ratio of the length of the side opposite the angle to the length of the hypotenuse. This mathematical function can be expressed with the following equation:

$$y = \sin x$$

Cosine (cos) θ $= \dfrac{A}{H}$

The cosine of the angle θ is the ratio of the length of the side adjacent to the angle to the length of the hypotenuse. This can be expressed with the following equation:

$$y = \cos x$$

Tangent (tan) θ $= \dfrac{O}{A}$

The tangent of angle θ is the ratio of the length of the side opposite the angle to the length of the side adjacent to the angle. This can be expressed with the following equation:

$$y = \tan x$$

How Tall is the Eiffel Tower?

Time to put all this into practice. Imagine that we want to figure out the height of the Eiffel Tower in Paris, France. Let's assume that we are standing 173 meters from the center of the base of the tower, looking up at the top, our head tilted at an angle of 60°. If we impose a right-angled triangle on top of these positions (because it's easiest to use), we find that the right-angle slots nicely into the base of the tower. Let's call the angle at the base of the tower angle B. We are standing at angle A and the top of the tower is home to angle C. From here we draw up the following equation:

$$\frac{BC}{AB} = \tan 60°$$

To find BC (the height of the tower), we arrange the equation to give the following:

$$BC = \tan 60° \times AB$$
$$BC = 1.73 \times 173$$
$$BC = 300$$

So now you know the Eiffel Tower is about 300 meters tall.

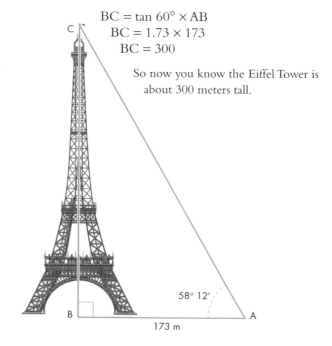

C

58° 12′

B

A

173 m

PHYSICS &
CHEMISTRY

Force and Motion

The poet Alexander Pope, arguably the most sarcastic Englishman in history, once wrote, "Nature, and nature's laws lay hid in night; God said, Let Newton be! And all was light" and for a while no one could have disagreed with him.

During the late seventeenth and early eighteenth centuries Newton effectively wrote the rule book on forces and motion, discovering the law of gravity and defining three laws of motion which we still use today. He also made important discoveries on the nature of light, designed the first reflecting telescope, and invented calculus.

Laws of Motion

Newton published his laws of motion in 1687 in a book called *The Mathematical Principles of Natural Philosophy*, which is generally known by the shortened version of its original Latin title, *Principia Mathematica*. Arguably the most important work of science yet written, it continues to provide a useful framework for examining force and motion.

Newton's First Law of Motion
Unless a force acts upon it, a stationary object will remain at rest and a moving object will continue to travel in a straight line at constant speed.

A good example of this first law is to consider a a ball lying still on a pool table. The ball will remain as it is unless it is struck with a cue or by another ball, or if someone tilts the table. Similarly, once in motion it will continue on its way until it is impeded, either by another ball, a side cushion or a pocket.

Newton's Second Law of Motion

The rate at which a stationary object moves from its position of rest, or the extent to which a moving object will deviate from its straight-line path, is dependent on the mass of the object and the force that is applied to it.

In other words, when a force acts on an object, the motion of the object will be affected.

For example, if we have a pool ball and a bowling ball (which has a greater mass than a pool ball) and attempt to push them forward with the same degree of force then inevitably the pool ball is going to move off more rapidly than the bowling ball. Once in motion, a greater force is needed to change the path of the bowling ball than that which is required for the pool ball.

Newton's Third Law of Motion

For every action there is an equal and opposite reaction.

Newton's third law of motion is perhaps the best-known of the three. It is known as the principle of **action and reaction**. It means that if one object exerts a force on another object then an equal force, called the **reactive force**, will operate in the opposite direction.

Perhaps the easiest way to illustrate this is to imagine a sprinter pushing away from starting blocks. As he or she pushes back against the blocks, an equal push is felt in the opposite direction, helping the runner off to a flying start.

Newton's laws of motion can be extended to apply to systems of particles and to continuous bodies by assuming that these are collections of particles.

Kepler's Laws of Planetary Motion
1. All of the planets follow elliptical orbits around the Sun, with the Sun as one focus of the ellipse.
2. A line drawn between the Sun and any of the orbiting planets sweeps out equal areas of the ellipse in equal lengths of time.
3. The square of the time taken for a planet to orbit the Sun is directly proportional to the cube of its mean distance from the Sun.

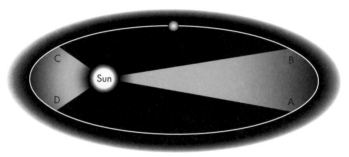

Planet in elliptical orbit around the Sun.

Based on observational evidence, Kepler formulated the above laws, the second of which had the greatest influence on Newton's thinking on gravity.

Newton's Universal Theory of Gravitation
The more massive the two objects are, and the nearer they are to each other, the greater will be the gravitational attraction between the two objects.

Building on Kepler's second law, Newton showed that the strength of the gravitational force between two objects is dependent on three factors: the mass of the first object, the mass of the second object, and their distance from each other. This relationship is governed by the **inverse square law**, which states that the attraction between two objects decreases with the square of the distance between them. For example, if the Earth were twice its current distance from the Sun, the magnitude of the gravitational attraction between the Sun and the Earth would be a quarter of what it is now.

Newton's Universal Law of Gravitation is stated in the following equation, where F is the strength of the force, m_1 and m_2 are the two masses, r is the distance between the centers of the two masses and G is a figure called the "universal gravitational constant."

$$F = \frac{Gm_1m_2}{r^2}$$

Newton's theory on gravity reigned until the twentieth century and the publication of Albert Einstein's theories of relativity.

Hooke's Law
1. In the case of relatively small deformations of an object, the size of the deformation is directly proportional to the deforming force.
2. All things remaining equal, the object will return to its original shape upon removal of the deforming force.

When Isaac Newton was a mere boy, the English scientist Robert Hooke discovered the law of elasticity which now carries Hooke's name. Its meaning in short is that if you don't squash an object too much it will spring back into shape.

According to Hooke's Law, the elastic behavior of a solid can be explained by small displacements of its atoms or molecules and that these displacements are proportional to the load placed upon them.

Hooke's Law can be expressed in the following equation, where the applied force (F) is equal to a constant (k) multiplied by the displacement (x).

$$F = kx$$

The Nature of Energy

Energy is a controversial topic among scientists. Everyone knows what energy does but no one seems to be entirely certain as to what it is. Essentially, there are two models used to describe the nature and behavior of energy: the **transformation model** and the **transference model**.

The Transformation Model

The transformation model has at its core the notion of "types of energy" and explanations based on this model will generally refer to "heat energy," "sound energy," "electrical energy," etc. Although very useful for introducing some of the basic concepts associated with energy this model has now fallen from grace.

The Transference Model

The transference model carefully avoids any mention of "types" of energy, preferring instead to accept the notion that energy simply exists and that it can be transferred, stored, conserved, or dissipated. Under this model, heat is not a type of energy but is instead a consequence of the transference of energy from one system to another.

Heat and Temperature

If we take the view that heat is a consequence of the transference of energy from one system to another then we can regard temperature as a measure of the amount of energy in a system before or after transference has taken place.

Temperature Scales

Any measurement of temperature is, for the most part, an entirely arbitrary process. Temperature scales take as their starting point two reproducible events and assign figures to them. In the case of the Celsius scale the two events are the freezing and boiling points of water. The freezing point is assigned a figure of 0°C and the boiling point is assigned a figure of 100°C. The space between these points, which is known as the **fundamental interval**, is then divided into degrees to give us the Celsius scale.

Another arbitrary temperature scale is the Fahrenheit scale, which, like the Celsius scale, uses the freezing and boiling points of water as its reproducible events but splits the fundamental interval, i.e. the resulting gap, into 180 degrees rather than 100 degrees.

International Practical Temperature Scale

Neither Celsius nor Fahrenheit is really suitable, however, for much of the work that is carried out by scientists and so the International Practical Temperature Scale was adopted instead. The current version

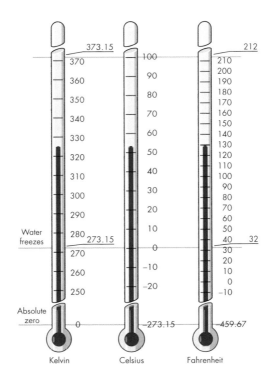

of this scale has 16 reproducible events (rather than the two of the Celsius and Fahrenheit scales), which are defined by thermodynamic temperature. The basic unit of thermodynamic temperature is the **kelvin**, named after the founder of modern British physics, Lord Kelvin (AKA Willy Thomson).

These 16 events are based on the triple points of various elements. (A triple point is the stage at which the temperature and pressure of the vapor, liquid, and solid phases of a substance are at equilibrium.)

Pressure

Pressure is a measure of the ratio of a force to the area affected. It is measured in units called **newtons** and is expressed as **newtons per square meter**. This can be stated using the following formula, where P is the pressure, F is the applied force, and A is the area over which the force acts:

$$P = \frac{F}{A}$$

Solids

The atoms or molecules of a solid are more densely packed than those of a liquid or a gas. When a solid is subjected to pressure it is said to be compressed. Temperature can also affect a solid, causing it to expand.

Liquids

The atoms or molecules of a liquid are less densely packed than those of a solid but more densely packed than those of a gas. Liquids exert pressure on any object immersed in them.

Gases

The atoms or molecules of a gas are less densely packed than those of either a solid or a liquid. Because their atoms and molecules are loosely arranged and in constant motion gases are easier to compress than either solids or liquids.

The Gas Laws

Gases behave in highly predictable ways, which have been described using three simple laws. These laws are collectively known as the gas laws and they describe the relationship between the temperature, pressure, and volume of a gas.

Boyle's Law
The volume of a mass of gas changes in relation to its pressure.

Named for the seventeenth-century Anglo-Irish chemist Robert Boyle the meaning of this law is that if the pressure on a gas is reduced, the volume of the gas (i.e. the amount of space it fills) will increase. Increasing the pressure on the gas has the reverse effect. Another way of expressing Boyle's Law is to state that the pressure and volume of a gas are inversely proportional to each other.

Charles's Law
The volume of a mass of gas that is at a fixed pressure is determined by its temperature.

Named for the French physicist J. A. C. Charles (1746–1823), the meaning of this law is that as the temperature increases, so too does the volume of the gas. A reduction in temperature has the opposite effect. Following on from this is the Pressure Law:

The Pressure Law
The pressure exerted by a gas at constant volume increases as the temperature of the gas rises.

Pressure and Depth

When an object is immersed in a liquid it is placed under pressure. Should the weight of the liquid displaced by the object be less than the object itself then the object will sink. As it sinks, it will experience an increase in the amount of pressure on it as a result of the weight of the liquid that lies above it. This pressure increases as the object sinks further into the liquid.

Upthrust

The pressure on an object immersed in liquid results in an upward force, called upthrust, which is equal to the weight of the liquid that is displaced by the immersed object. If the force of upthrust acting on the immersed object is greater than the weight of the object itself then the object will float. If it is less then the object will sink.

Fluid Dynamics

All observations on the nature of liquids and gases can be brought under the general banner of fluid dynamics. This is the study of the effects of forces and energy on liquids and gases. For the purposes of this branch of classical physics, liquids and gases are treated as essentially the same thing (fluids) and are investigated using the same equations. Fluid dynamics can be applied to investigations in hydraulic, aeronautical, and chemical engineering.

Electromagnetism

Electricity and magnetism are two parts of the same fundamental force, called electromagnetism. Magnetism can be used to generate electricity and electricity can be used to create magnetic fields.

Electricity

An **electric current** is created by the flow of electrons. Electrons can move easily between the atoms of some metallic elements, such as copper. As they do so, they repel electrons orbiting the nucleus of the atom to which they have moved, causing them to move on to the next atom, and so on. This flow of electrons over the short distances between atoms carries on through the conducting material (unless opposed), creating what we think of as an electric current.

Type of current	Definition
Direct current (DC)	The flow of electrons in a single direction
Alternating current (AC)	A flow of electrons that alters direction many times per second

Amps
Electric current is measured in units called amps. One amp is equal to the flow of six million trillion electrons per second.

Volts
The force that powers and drives the flow of electrons is called the **electromotive force**. It is measured in **volts** and the amount of electromotive force in a circuit is called the **voltage**.

Resistance
Some materials conduct electricity better than others. The degree to which a material opposes the flow of electrons is its resistance, measured in **ohms**, the symbol for which is "W."

Ohm's Law
1. The current is equal to the voltage divided by the resistance.
2. The voltage is equal to the current multiplied by the resistance.

Named after German physicist Georg Simon Ohm, this law is concerned with the relationship between current (I), voltage (V), and resistance (R). This is expressed in a simple equation:

$$I = V \div R \quad \text{or} \quad V = I \times R$$

Electric Circuits

When components and a power source are combined for a purpose, such as to provide a lighting system for a house, they are arranged in a circuit. There are two basic types of electric circuit: the **series circuit** and the **parallel circuit**.

Series Circuit

In a series circuit the power source and the components are arranged one after another in a loop. Current flows through each of the components in turn but resistance in the circuit reduces the effectiveness of the current. Any break in the circuit will result in its complete breakdown, with current lost to all components.

Parallel Circuit

In a parallel circuit the components are arranged on separate "branches," which effectively means that they are each connected directly to the power source. This reduces the amount of resistance that is "felt" in the circuit and also means that any break in one branch of the circuit can leave the rest of the circuit unaffected.

Electromagnetic Field

Magnetism, like electricity, is caused by the movement of electrons inside atoms. The atoms of some materials, such as iron or cobalt, are much more magnetic than others. These are known as **ferromagnetic** materials. As the electrons spin inside atoms they generate very small magnetic fields. When atoms of ferromagnetic materials are placed together they reinforce each other, creating much larger electromagnetic fields. These "mini-magnets" are grouped together in small areas called domains. The alignment of these domains will determine whether or not the material will attract or repel other ferromagnetic materials.

Generator

Back in the first half of the nineteenth century the English scientist Michael Faraday discovered that if he moved a coiled wire through a magnetic field he could generate an electric current inside the coil. This discovery formed the basis of what is now known as *Faraday's Law of Electromagnetic Induction* and remains the process by which electricity is generated to this day.

Electric Motor

At a basic level the electric motor is simply a coil of wire held on a spindle between the poles of a magnet. Passing a current through the coil effectively magnetizes it, causing it to spin on its axis as it is repelled and attracted by the surrounding magnet. To ensure that the coil moves in one direction only, a device called a **commutator** switches the direction of the current with every half turn of the coil so that it is pushed up on one side and pulled down on the other in a continuous movement. By attaching a shaft to the spinning coil one can tap into this movement and use it to drive anything from an electric toothbrush to a subway train.

Electromagnetic Spectrum

The electromagnetic spectrum is the range of wavelengths over which electromagnetic radiation extends. It goes from radio waves (the longest waves) at one end of the spectrum up to gamma rays (the shortest waves) at the other.

Elements

It's a useful cliché to note that everything on Earth (and beyond) is made up of atoms of one or more elements. An element is a substance, such as iron, oxygen, or gold, that consists entirely of identical atoms.

Atoms

The atom is the smallest part that an element can be divided into. It generally comprises a positively charged central core of protons and neutrons, called the nucleus, which is orbited by negatively charged electrons. The electrons determine the chemical behavior of atoms, dictating the way in which the atoms will bond with other atoms, be it by the sharing or transferring of electrons.

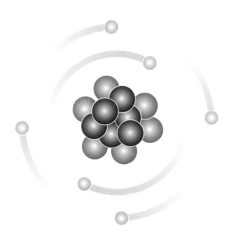

Carbon atom with protons and neutrons in the nucleus and six electrons in orbitals.

Ions

Atoms normally have neutral charges but electrons can move between atoms during chemical reactions causing them to gain an overall positive or negative charge. These electrically charged atoms are called ions. A positively charged ion is called a **cation** and a negatively charged ion is called an **anion**.

Molecules

A molecule consists of two or more atoms that are chemically bonded – these can be atoms of the same or different elements. Atoms bond to each other to become molecules by "sharing" electrons.

A molecule is the smallest amount a substance can be broken down into and still retain its basic properties. For example, the water molecule (H_2O) is made up of two atoms of hydrogen plus one atom of oxygen. In this form it has all the properties we associate with water – liquid at room temperature, boiling at 100°C, frozen at 0°C, etc. But if we separate the hydrogen and oxygen atoms then we lose the water molecule and are left with two elements that are gases at room temperature.

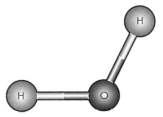

Water molecule: H_2O

The composition of any molecule of a substance can be worked out by careful examination of its chemical formula. We've already seen that water (H_2O) is made up of two hydrogen atoms and one oxygen atom. But what about a molecule with the chemical formula H_2SO_4

If you refer to the Periodic Table over the page, you'll see that the atoms in the molecule are those of hydrogen, sulfur, and oxygen. Working from left to right, we can clearly see that the molecule comprises two hydrogen atoms, one sulfur atom, and four oxygen atoms. These elements combine – in these proportions – to form a molecule of sulfuric acid. If we could remove the sulfur atom and three of the oxygen atoms we would be left with water.

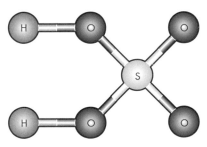

Molecule of sulfuric acid: H_2SO_4

Compound

A compound is a substance formed by the combination of elements in fixed proportions.

The Periodic Table

Created by the Russian chemist Dmitri Mendeleyev in 1869, the Periodic Table is a visual arrangement of elements according to their reactive properties and their atomic number (which is determined by

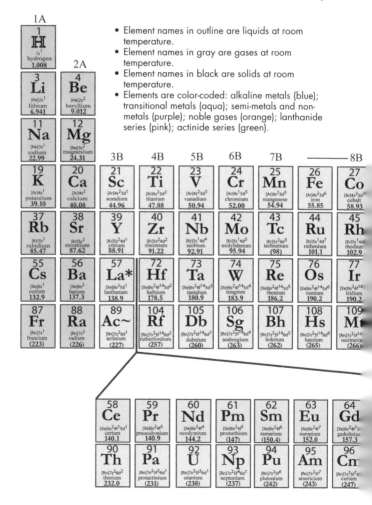

- Element names in outline are liquids at room temperature.
- Element names in gray are gases at room temperature.
- Element names in black are solids at room temperature.
- Elements are color-coded: alkaline metals (blue); transitional metals (aqua); semi-metals and non-metals (purple); noble gases (orange); lanthanide series (pink); actinide series (green).

the number of protons in the nucleus of the element). It is arranged in vertical columns called **groups** and each of the horizontal rows is known as a **period**.

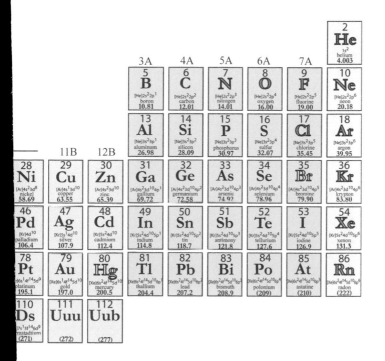

Einstein's Theories of Relativity

Albert Einstein became the poster boy for science and mathematics after his thoughts on relativity were brought to the attention of the masses in 1919. The world's press hailed his theory and the apparently successful proof of it.

In reality, few people had any idea what the theories were about and those who did had difficulty believing them. In truth, relativity isn't so much hard to understand as hard to believe because at its core was the conclusion that the Universe didn't really work the way that most of us – including Sir Isaac Newton – imagined it did.

Special Theory of Relativity

Einstein published his Special Theory of Relativity in 1905. Why was it termed "special"? Because it is concerned with the behavior of moving objects in the "special" environment empty space free from gravity.

He started with the premise that the one absolute in the Universe is the speed of light and that everything else, including time and space – or "space-time" as Einstein preferred to think of it – is relative to one's frame of reference. Working from this point he then made some truly alarming but entirely fascinating predictions about the nature of mass, energy, and time.

Mass–Energy Equivalence

Chief among Einstein's predictions is the idea of mass–energy equivalence. This is to say that mass and energy are essentially the same thing and that one can be converted into the other. This is expressed in the famous equation $E = mc^2$.

Energy and Mass at the Speed of Light

Einstein then added that as a body (yours, for example) accelerates through space its energy and mass increase while its length decreases in the direction of travel. In short, the faster you go the "heavier" you get and the shorter you become.

As you approach the speed of light this effect increases and if you could reach light-speed your body would have infinite mass and zero length. This would, of course, be impossible but it does illustrate the point that the speed of light is the top speed attainable in this Universe.

Time Relative to Observer's Frame of Reference

The final prediction suggested by the theory is the one that most people have trouble getting their heads around. It suggests that as your body is hurled through space, time appears to pass at a normal speed for you but to an observer who is outside your frame of reference, time will appear to slow down.

We don't notice this at normal highway speeds but at near-light speeds it becomes more obvious. This suggests that even time itself, like space, is not an absolute but does in fact pass at a speed that is relative to the observer's frame of reference.

General Theory of Relativity

According to Sir Isaac Newton, gravity is a force of attraction between two bodies that is felt instantaneously across the vast reaches of space. Unfortunately, this is incompatible with the Special Theory of Relativity, which states that nothing moves faster than light. Einstein addressed this problem with the General Theory of Relativity, which he published in 1915.

According to the General Theory of Relativity, a gravitational field is the result of space-time bending around a body with mass. This is easier to picture if you think of what happens when you sit on a mattress. As you sink your weight into the bed the mattress bends around you and a you-shaped dip is created. Anything that is already lying on the mattress will inevitably fall in your general direction.

In a similar way, our Sun puts a big dent in the fabric of space-time, causing it to curve and distort. Anything passing near it, such as a planet, will inevitably follow a path that is dictated by these curves and distortions. According to Einstein, this, rather than some mysterious and instantaneous force of gravity, is what causes gravitational "effects." In short, what we think of as gravity is actually a consequence of the bending of space-time.

$E = mc^2$

There's no denying that the most famous equation in human history is known to an awful lot of people, but very few know what it is actually telling us.

First off, we need to define terms. $E = mc^2$ states that energy (E) is equivalent to a unit of mass (m) multiplied by the constant speed of light (c), squared (^{2}).

Since the square of the speed of light is 90,000,000,000,000,000 meters per second we can be certain that a figure multiplied by this amount is going to be vast and that liberating the energy quickly from even a small amount of mass is going to be a dramatic event – this is why a nuclear bomb containing only 100 grams of convertible mass (about the same as you'd find in a small jar of instant coffee) can destroy an entire city.

Despite Einstein's connection with the theory that lay behind the development of the nuclear bomb, he had no direct involvement in its construction. Although he had urged the U.S. to investigate the possibility of building such a weapon (on the grounds that there was evidence to show that some of his former colleagues were helping the Germans to do likewise), he was horrified when he discovered that it had been used against civilian targets. The experience made him a leading advocate of world peace.

FACT

When Einstein arrived in America to take up a teaching post at Princeton University in 1934 the press was there to welcome him and the story made the papers and the newsreels. As a consequence, Einstein spent the rest of his professional career fielding requests for help with homework from enterprising (and very cheeky) American schoolchildren.

Radioactivity

Unstable atomic nuclei emit high-energy subatomic particles which we commonly call **radiation**. Some nuclei do this naturally while others do so after being bombarded by subatomic particles (see pages 84–85).

There are three types of radiation – **alpha** (α); **beta** (β); and **gamma** (γ) – which are classified according to their ability to penetrate matter. Often thought of as "rays" of radiation, in fact only gamma radiation is emitted in the form of rays; alpha and beta radiation have been found to be made up of streams of subatomic particles.

Changes to the nuclei cause the emission of alpha and beta radiation. This is known as **radioactive decay**. Gamma rays are a consequence of the nucleus throwing off excess energy. All three forms of radiation are potentially harmful to humans.

Properties of the Alpha Particle

Alpha particles have two protons, two neutrons, and a positive charge. They are unable to penetrate the skin but if inhaled or swallowed can trigger changes at the cellular level which will most likely cause cancer.

Properties of the Beta Particle

A beta particle is an electron that is emitted when an excess neutron in the nucleus of an atom changes into a positively charged proton, causing the electron to fly off, taking the excess negative charge with it. Beta particles can penetrate the skin but are more likely to be inhaled or swallowed. Depending on where the beta particles end up, they can trigger cancers such as leukaemia.

Properties of the Gamma Rays

Gamma rays are similar to X-rays and are a form of electromagnetic radiation. They are to be avoided as not only can they penetrate the skin, they can also penetrate lead and concrete. They create charged atoms called ions on contact with the human body, which cause living tissue to break down.

Fission, Fusion, and the Bomb

As we have already seen in our look at the Special Theory of Relativity (page 80), mass and energy are two faces of the same coin. Matter, or mass, is converted into energy in one of two ways: **nuclear fission** or **nuclear fusion**. Any conversion of matter releases enormous amounts of energy, which can be used for good or for ill.

Nuclear Fission

Nuclear fission relies on the fact that isotopes such as uranium-235 and plutonium-239 become highly unstable when they pick up extra neutrons. This causes them to split in an instant, producing more neutrons, which then go on to destabilize further nuclei in a process known as a **chain reaction**.

Energy, in the form of gamma radiation (among others), is released, producing enormous amounts of heat. Nuclear reactors are designed to ensure that this is a controlled process, with just enough neutrons released to keep the chain reaction going. With a nuclear bomb, the chain reaction is designed to occur in a few millionths of a second with the result that the energy release is truly explosive.

Nuclear Fusion

Nuclear fusion requires the heat of a fission reaction and uses it to fuse light nuclei together. In the Sun, this occurs when the nuclei of hydrogen fuse to form helium, releasing huge amounts of energy. Although the Sun supplies almost all of the energy in our Solar System we have yet to build a reliable fusion reactor here on Earth.

FACT

The Manhattan Project saw the world's first nuclear bomb developed in the U.S.A. When the time came to test the weapon, some scientists expressed the fear that it might trigger a chain reaction that would destroy the world. But they set the bomb off anyway.

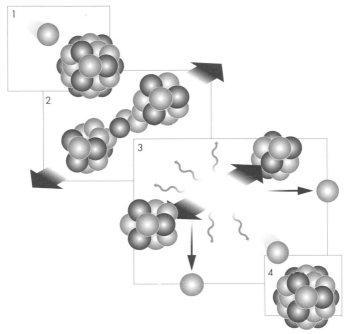

Nuclear Fission
1 Uranium-235 nucleus is struck by neutron. It absorbs neutron and becomes uranium-236.
2 This splits, forming two similar nuclei.
3 The split releases radiant energy and releases several more neutrons.
4 These neutrons start the process again, splitting other nuclei.

This is a shame because a commercial fusion reactor could provide an almost limitless source of electricity. Unfortunately, the problems associated with the costs, combined with the degree of difficulty involved in building such a reactor, have so far proved insurmountable.

Fusion experiments have been carried out on a smaller scale, however, suggesting that some form of fusion-based electricity generator may be feasible within a generation.

Subatomic Particles

To the best of our knowledge, all of the elements in the Universe are made up of atoms. The atom is the smallest unit to which an element can be reduced but in the last hundred or so years we have discovered that the atom itself is made up of smaller, i.e. subatomic, particles.

Electrons

The most interesting of the subatomic particles is undoubtedly the electron. The electron is the smallest of the three principal subatomic particles. It is a negatively charged particle of energy that races around the nucleus of an atom at the speed of light in an irregular pattern known as an **orbital**. (An orbital can be thought of as a shell around the core of the atom.) By taking in energy, electrons can leap to a higher shell, or drop to a lower one if energy is lost. Because electrons behave like fuzzy clouds of negative charges around the core of the atom it is impossible to predict their exact position at any given point in time.

Protons

If it were possible to journey to the core of an atom one would find a tightly packed nucleus. This houses most of the mass of the atom and generally consists of protons and neutrons (the hydrogen atom has just a single proton).

Protons have a mass that is 1,836 times that of electrons. They also have positive electrical charges which balance an equal number of negatively charged electrons, making the atom electrically neutral. It is this electromagnetic attraction between electrons and protons that holds the atom together.

FACT

If an atom's protons and neutrons were the size of tennis balls, the electrons would be the size of pinheads and the whole atom would extend over several thousand meters.

Neutrons

Neutrons, as the name suggests, have no electrical charge. Their mass is 1,839 times that of the electron. Like the proton, they can be found at the core of the atom.

Quarks

The protons and neutrons that make up the core of an atom are themselves made up of three smaller particles called quarks. There are two types of quark found in protons and neutrons: the **up quark** and the **down quark**.

String Theory and Quantum Loops

Particle physicists have been having a ball recently. It seems that barely a day goes by without them discovering some new subatomic particle or other. At present over 200 types of subatomic particle are thought to exist, although the evidence for some of these is sketchy at best.

Particles such as the electron, which were previously not thought to be made up of smaller components, are said to be elementary particles. However, some physicists believe that even these particles are made up of quantum loop-type units called **superstrings**. These are thought to be billions of times smaller even than the fundamental particles and are believed by many physicists to form the real fabric of the Universe. If true, this would provide us with a theory that could explain all of creation (but don't hold your breath).

A Brief History of Cosmology

Cosmology is a branch of astronomy that examines the really big picture. It's concerned with the origins, nature, and scale of the Universe and seeks to answer some of our oldest questions, such as "How did we get here?" "Where are we going?" and "How long have we got left?"

Cosmology might be said to have been "invented" some time around 3000 B.C. when the Babylonians made the first systematic study of the night sky, identifying several constellations in the process.

The Greeks joined in some time around 400 B.C. with Aristotle realizing that the Earth is a sphere and Eratosthenes working out a roughly accurate figure for the diameter of the Earth.

Ptolemaic Universe

By the second century A.D., Ptolemy (with a silent "P") proposed a model of the Universe that placed the Earth at its center. Nonsense, of course, but this theory remained largely unchallenged until the sixteenth century, when Nicolas Copernicus pointed out that the Earth and the other planets orbit the Sun. Galileo stole this idea, found proof for it, and almost got himself burned at the stake for his trouble.

Heliocentric Universe

Improvements to the design of the telescope in the seventeenth and eighteenth centuries vastly improved our understanding of the Universe and by the twentieth century we were finally beginning to realize that we live on a small rock that orbits the Sun in a truly vast and mostly hostile Universe.

Hubble's Law

American astronomer Edwin Hubble was the first person to get a glimpse of how truly enormous the Universe is when he began studying what was then called the **Andromeda Nebula**. While attempting to figure out how far away the nebula is, he realized to his surprise that not only is it too far away to be part of our Milky Way Galaxy (two million light years and counting) but that it is also a galaxy in its own right.

He then turned his attention to other nebulae and was no doubt thrilled to discover that many of these are also galaxies, located so far away that we had to rethink our notions of the scale of the Universe.

In the late 1920s Hubble began analyzing the light emitted by stars in these distant galaxies. Against expectation, he found that the wavelength of the light was shifting to the red end of the spectrum.

This effect, called **Red Shift**, indicated that the galaxies were moving away from us at a tremendous speed. This suggested that the Universe is expanding. And as if that wasn't enough, Hubble also found that the further away a galaxy is from us, the faster it will be moving. This discovery became known as Hubble's Law.

Hubble's Constant

Hubble's discovery also produced the Hubble Constant, which is the ratio of the speed of a receding galaxy to its distance from us.

Big Bang Theory

There have been many differing theories on the origins of the Universe but the only one that is currently accepted by the majority of astronomers is the Big Bang Theory. This states that the Universe – which is to say all matter-energy and space-time – burst into being when a quantum singularity split apart in an enormous explosion around 15 billion years ago. Within a second the Universe had grown in size by two billion, billion kilometers and is still expanding to this day.

The theory was first suggested in 1927 by a Belgian astronomer and priest called Georges Lemaître. He believed that the Universe began life as a dense, egg-shaped object about 30 times larger than the Sun, which he called the "primal atom Cosmic Egg." The theory certainly appeared to fit with the evidence for an expanding Universe that had been discovered by Edwin Hubble, although Lemaître's "primal atom Cosmic Egg" was far too large. But where was the evidence for the Big Bang itself?

Physicist George Gamow predicted that microwave radiation from the Big Bang should still be reaching us from the furthest depths of the Universe. In 1964, Arno Penzias and Robert Wilson discovered and identified this background radiation, and in 1992 a microwave map of the sky was produced by the Cosmic Background Explorer (COBE) probe, providing further evidence in support of the Big Bang Theory for the origin of our Universe.

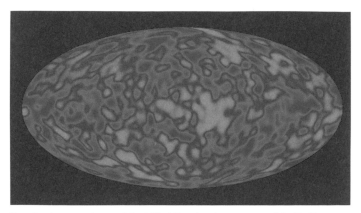

Cosmic microwave image of the Milky Way, showing remnants of the Big Bang.

Steady State Theory

For several decades the Big Bang Theory of the origin of the Universe faced competition from an alternative proposition called the Steady State Theory. This theory, first proposed by British astronomer Fred Hoyle, states that the Universe had no beginning and will have no end. Furthermore, despite its expansion the density of the Universe remains constant by virtue of the creation of new matter at the rate of one hydrogen atom per liter of space every 20 years (a figure that could not possibly be measured in a lab). The Steady State Theory received a mortal blow with the discovery of cosmic background radiation (a remnant of the Big Bang) in 1964.

FACT

Fred Hoyle, British astronomer and original proponent of the Steady State Theory, first used the expression "Big Bang" in a series of lectures in 1950 in which he mocked the idea of an almost biblical beginning to the Universe.

FACT

You can hear remnants of the Big Bang every time you tune your radio. Some of the hiss and crackle you can hear between radio stations is actually an echo of the Big Bang.

Brane Theory

Brane Theory ("Brane" being short for "membrane") is a whole new school of thought regarding the nature of the Universe. It grew out of a desire to explain the nature of gravity at the quantum level.

Of the four fundamental forces in the Universe (see box below), gravity is by far the weakest. This fact prompted some scientists to ask the question: "Is gravity really that weak or does it merely appear to be so because it is operating across dimensions that we cannot perceive?"

If we assume that a membrane is a two-dimensional surface separating two three-dimensional volumes, then according to Brane Theory the four dimensions that we can see (three of space and one of time) may actually be a membrane between volumes that contain additional, higher dimensions in space. These higher dimensions are inaccessible to us because we and the known fundamental forces (with the exception of gravity) are trapped in the membrane.

If true (and it's a very big "if"), then it would explain why gravity appears to be such a weak force. In essence, it is being diluted by having to operate over perhaps 11 dimensions – the four we know of plus another seven that are thought to exist.

Four Fundamental Forces of the Universe

1. Electromagnetic interaction
2. Gravitational interaction
3. Strong interaction
4. Weak interaction

Matter Fields

There is a certain amount of circumstantial evidence in support of Brane Theory, some of which points to the possibility that the Universe may have begun with the interaction of several membranes. There has been a lot of talk of **dark matter** in recent years. This actually appears as a gravitational force pulling matter together where no matter appears to be present. Yet this mysterious material may make up more than 90 percent of our Universe.

Brane Theory proposes that these matter fields may in fact be the manifestation of another membrane, which has influenced the formation and structure of our Universe and which, at the local level, is helping to hold our Milky Way Galaxy together.

Infinity Redefined

The Universe was at one time thought to be infinitely large and infinitely old. If we accept the Big Bang Theory of the origin of the Universe then we know that it must have had a beginning. As for the Universe being infinite in size, there is the small matter of **Olbers's Paradox** to contend with. This states that if the Universe is infinite then we would see starlight coming from every point in space, but we don't.

Part of the beauty of the night sky is in the way that the stars are laid out against a midnight blue background. Clearly, if there were an infinite number of stars in the sky we would not have the dark background.

Albert Einstein redefined the Universe when he stated that rather than being infinite it is in fact finite yet unbounded. To get your head around this concept try to imagine that the Universe is like the surface of a gigantic sphere. You could travel in any direction for as long as you liked but you would still be on the finite yet unbounded surface of the sphere.

COMPUTERS &
DIGITIZATION

Calculating Machines

Abacus

The abacus was the first true calculating machine and was probably invented by the Chinese some time around 500 B.C. The principle behind the abacus is simple. Working in base 10, beads are used to represent numbers and their value is determined by the line or row on which they are placed.

Starting at the top, the beads on the upper row represent single units, those on the row below are worth 10 times more, the beads on the row below that are worth 10 times more again (which means that a single bead is worth 100), and so on.

An alternative design of abacus used markers on a chequered cloth but in every other respect resembled the bead and row style of the abacus.

Example of Pascal's Calculating Machine from the seventeenth century.

Pascal's Calculating Machine

The invention of the first mechanical calculating machine some time in 1623 is credited to one Wilhelm Schickard, a friend of the astronomer Johannes Kepler. Unfortunately, we have no surviving examples of this machine and so we tend to view the adding machine built by the French scientist and philosopher Blaise Pascal between 1642 and 1645 as the first true mechanical calculator. This was a device that used cogs and wheels to perform addition and subtraction over eight columns of digits.

Leibniz's Calculating Machine

Gottfried Leibniz, of calculus fame (see page 55), came up with an improved calculator in 1673. Leibniz's machine was considerably more sophisticated than Pascal's and, as well as performing addition and subtraction, could be used to multiply, divide, and find square roots.

Jacquard's Loom

The next significant step on the journey towards the creation of the modern computer occurred in 1804 when the Frenchman Joseph-Marie Jacquard invented an automated loom. Patterns in cloth woven on the loom were dictated by a series of punched cards. This was the first time that controlling data had been stored on cards and then processed later in a machine.

Analytical Engine

Just 30 years after the invention of the Jacquard Loom, an English inventor named Charles Babbage conceived the world's first digital computer. Called the Analytical Engine, this was the first device designed to perform arithmetical calculations and then make simple "decisions" based upon these calculations. This design was extraordinary in that it combined, for the first time, most of the elements of a modern computer: a central processing unit, a rudimentary memory, a data input/output system (Jacquard's punched cards), and sequential control. Unfortunately, Babbage's ambitious design was frustrated by the limitations of the technology of the period – and a woeful lack of funds.

Boolean Logic

While Babbage was busy waiting for technology to catch up, another Englishman, named George Boole, was at work creating a new kind of algebra that would define logic as a branch of mathematical theory. Boolean logic used simple operators – "AND," "OR," "NOT" – and the binary system to produce, in 1847, a "language" that could almost

EXAMPLE OF BOOLEAN LOGIC IN PRACTICE

1 AND – used to locate instances where all terms are present

Example search statement: oil AND water AND pollution

Records retrieved: Documents that discuss all three terms oil, water, and pollution

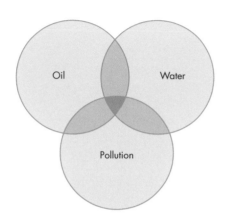

2 OR – used to locate instances where any one of the terms being searched for is present

Example search statement: cream OR cheese OR milk

Records retrieved: Documents that discuss one or more of cream, cheese, or milk

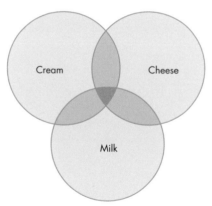

3 NOT – used to locate instances where the first term is present but the second is omitted

Example search statement: magazine NOT newspaper

Records retrieved: Documents that discuss magazines but not newspapers

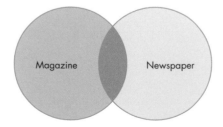

have been custom-written for the transistor circuits of the modern digital computer – quite some feat considering that this was exactly 100 years before the first transistor was developed.

By now the world stood on the verge of the computer age with most of the basics in place. There was a long series of digital switches capable of carrying through the binary logic necessary for complex mathematical calculations. Now all that was needed was a way of doing it electronically and on a manageable scale.

First Electronic Computer and Transistor

The first electronic computer was constructed at Bletchley Park in England in 1943. Designed to help crack German codes during the Second World War, this machine was built using valves and as such was enormous, very hot and none too reliable. Fortunately, four years later, in 1947, Messrs Bardeen, Brattain, and Shockley of Bell Laboratories in the U.S.A. developed the first transistor.

Computer Age

The transistor gave birth to the computer age. Smaller, trendier, and ever more powerful computers could be built using the simple switching abilities of the transistor. For the first time, people started to talk about the coming "Computer Age" and by 1951 the world's first commercially available computer, the Ferranti Mark 1, was put on sale. (The fact that this design was valve-driven may have had something to do with the company selling only eight of them.)

From 1958, people were building computers with transistors and for the first time the computer crossed over to the business world, having previously been the preserve of government offices and academic institutions.

BASIC Code

The Digital Equipment Corporation introduced the first mini-computer in 1963. BASIC (Beginner's All-purpose Symbolic Instruction Code) appeared in 1964 and suddenly there was a useful language available, making computer programming easier.

Integrated Circuit Chip

The mouse arrived in 1968 and the integrated circuit (IC) chip arrived two years later. This saw machines get even smaller and more powerful – the first Intel microprocessor chip followed in 1971. The IC chip also aided the introduction of the first personal computer, the Altair 8800, which was offered for sale (in kit form!) in 1975. Although not a huge success, it arguably gave IBM the shove they needed to produce their own personal computer in 1981.

Graphical User Interface

Three years later Apple launched the Macintosh range which, significantly, made extensive use of the Graphical User Interface and thus brought computing into the realm of those who previously would have dismissed computers as toys for geeks. So successful was this interface – almost anyone could use it – that it forced other computer manufacturers to attempt to adopt this user-friendly approach.

Commercial Computer Sales

By 1998 Bill Gate's Microsoft was the most valuable company in the world, a reflection, surely, of just how far things had come in an astonishingly short time period. Less than 40 years earlier a total of eight computers had been sold on the commercial market. Today the computer plays a significant role in almost all aspects of our lives, a situation that doesn't look likely to change in the near future.

Computer Components

There are lots of components in the modern microcomputer, but for the most part they can be divided into two categories: **storage** and **processing**. Should you decide to take the back off your home computer, you'll find all of the components that are described on the opposite page in some form or other (and possibly discover that you've invalidated your warranty). If you do remove the cover, be careful not to touch anything as even a small static discharge can destroy a delicate circuit.

Component	Function
Central Processing Unit	Device that handles data processing. It controls all of the input, output, and storage devices. In most computers the CPU is easy to spot because, being as busy as it usually is, it generates enough heat to need a dedicated fan.
Motherboard	Main circuit board where all of the important components of the computer are wired together.
Hard-disk Drive	Metal disk or disks where data is stored permanently on the computer in the form of magnetic signals arranged in binary code. Magnetic heads, much like those used to read videotape, write data to the disk and can read it off again when needed.
RAM (Random Access Memory) Chip	Chips that provide a temporary home for data and applications that are currently in use. Any information held on these chips is lost when the computer is turned off if it hasn't already been transferred ("saved") to the hard-disk drive.
BIOS Chip	This chip carries essential information such as the boot-up sequence needed to start the computer. (Don't touch it!)
Controller Chips	Chips designed for a specific purpose, such as running complex graphics or digital video packages.
CD/DVD Drive	Data storage drive designed to allow information to be read or stored on removable media such as floppy disks, CD-R/RW, and DVD.
Adder Circuits	Circuits that use binary logic to allow addition, subtraction by the addition of a negative number, multiplication by using repeated addition, and division by repeated subtraction.

FACT

Tuvalu, a remote island group in the southwestern Pacific Ocean, sold the rights to its domain suffix (.tv) for 50 million dollars. The rights revert to the island after 10 years.

Alphanumeric Characters

Every computer that uses the Roman alphabet, i.e. most of them, also relies on a standardized binary code called ASCII, which stands for "American Standard Code for Information Exchange." In use since 1963, this has been the standard machine code for computers ever since.

Like other codes before it, ASCII uses binary numbers to represent letters, numbers, and other symbols. It employs a string of eight binary digits or "bits" to represent 128 characters. These characters are enough for all of the letters of the Roman alphabet, every number between 0 and 9 (and their various permutations), all the common punctuation marks, and 32 special characters used to control computer functions. An additional 128 characters are used for all of the other stuff, including accented letters and the occasional symbol, such as this ©.

You can get a sense of just how fast computers work by considering the time taken between pressing the dollar sign on your keyboard and its appearance on screen. When you hit "$" on the keyboard it generates the ASCII code 36. This is immediately converted into its binary equivalent (00010010), which is then processed by your computer, whereupon the symbol appears on screen. See page 139 for more on computer coding.

FACT

Languages such as Chinese, which are not based on the Roman alphabet, require the use of double the number of binary digits in computing to achieve similar results.

Digital Sound

Digitization is not restricted merely to letters, numbers, and a few special characters. These days we can digitize just about anything in order to bring it onto our computers. Sound was first digitized as part of ongoing improvements to the telephone system. When sound is converted from its original analog form into a digital format it is first "sampled." This involves "measuring" the signal at regular intervals and giving it a binary number.

Telephone systems, which need to handle only the relatively limited range of the human voice, rely on an eight-bit system to do this. Each sampling of the analog sound wave produces an eight-digit binary number. (By the way, an 8-bit system samples the sound wave 8,000 times per second.) Each and every sample is given one of 256 different values (256 being the highest number achievable with eight binary digits). This can then be transmitted and reproduced at the other end of the line where a digital-to-analog converter turns the digital signal back into speech.

Much higher sample rates are used for music-based technology, with CDs sampling at 44,100 times per second and DVD-Audio achieving a truly awesome figure of almost 17 million samples per second. (And no doubt the last sentence will seem comical 10 years from now.)

Digital Pictures

Digital images are created by scanning a picture in very narrow strips (a little like ploughing a field, but on a considerably smaller scale). These strips are then further divided into small squares called **picture elements**, or **pixels**. Each pixel is assigned a code, which reflects the amounts of red, green, or blue it contains. The brightness of the pixel is also measured and given a binary number.

The result of all this scanning and coding is a **bitmap**, i.e. a map of the bits of information that make up the image, which can be stored, transmitted, or reproduced on a screen or as a printout using the appropriate technology.

The quality of the resulting images depends almost entirely on the amount of information that was scanned and stored in the first place. The deciding factors are the number of pixels in a given area and the number of levels of brightness against which each pixel is measured.

Typically, the highest-quality color images have 22,500 pixels per square inch, which will have been sampled at 12 bits-per-color. This degree of quality and accuracy is about eight times better than that of a normal computer or TV image.

Digital Video

Digital video, which combines the elements required for digital sound and digital images, first appeared as a rather rough-and-ready format. Images stored on cinema film are typically shown at a rate of 24 frames per second. By comparison, early digital video ran at five frames per second and looked crude and grainy. Improvements in storage capacities and processing speeds mean that, in terms of quality, digital video can rival and even surpass that of film. Combine this with CD-quality sound and it's easy to see why the digital revolution has taken hold of our imaginations.

Unit of storage	Definition
1 bit	Single binary digit (either 0 or 1)
1 byte	8 adjacent binary digits (8 bits)
1 kilobyte	1,024 bytes
1 megabyte	1,048,576 bytes
1 gigabyte	One billion bytes
1 terabyte	One trillion bytes

Data Storage

From the very first floppy disks to today's multiple-array hard drives, the capacity of data storage devices has grown at such an astonishing rate that to give a figure for the device with the largest capacity would be to invite mockery a year or so from now. But to give some idea of the rate of growth, consider this: in the 1970s the now almost-obsolete floppy disk held a maximum of 1Mb of data; in the 1980s a CD-ROM could hold about 700 Mb of data, while a single DVD from the 1990s is capable of storing 25 times as much information as a CD.

New data compression techniques will see a growth in the capacity of data storage devices that might sensibly be compared to the improvement in the speed of communication that came with the introduction of the telegraph.

Moore's Law

Many people who bought a computer perhaps five years ago and then upgraded it in the last year or so may have been surprised to discover that they paid roughly the same amount on each occasion. They will, however, have got a considerably faster and more sophisticated computer the second time around. This is reflected in something known as Moore's Law.

Named for Gordon Moore, an executive at Intel, Moore's Law states that microprocessors double in complexity and power every two years. Starting with the Intel 80286 microprocessor, which was introduced in 1982, we can see that after a slow start, if anything, Moore underestimated the improvement in microprocessing power that has occurred since then. (Microprocessor "clock" speeds are measured in **megahertz** and, more recently, **gigahertz**.)

Processor	Clock speed	Year of launch
Intel 80286	8MHz	1982
Intel 80386	16MHz	1985
Intel 80486	25MHz	1989
Intel Pentium	60MHz	1993
Intel Pentium II	350MHz	1997
Intel Pentium 4	1,500MHz	2000

Computers and Science

The development of the computer could not have come quickly enough for the twin worlds of science and mathematics. Suddenly alarming amounts of research data could be processed in a fraction of the time previously needed for such things. At a stroke astronomers were, for example, able to digitize images of large portions of the night sky taken at the same time on different nights and leave the computer to search for any signs of deviation (which might highlight the whereabouts of comets, supernovae, and other items of interest).

Biologists are able to compare DNA samples across a broad spectrum of society while forensic scientists and the police can find matching fingerprints from a huge pool of samples. Physicists and chemists can build virtual models to test their theories and mathematicians have been able to find π to over a billion decimal places.

Virtual Worlds

Of all the benefits to science and society that the modern computer has brought, it is perhaps in the field of concept modeling that the computer has proved itself to be an invaluable tool. By programming in all of the known factors that might affect, for example, the stresses and strains to which a particular bridge might be subjected one can then try out the design in a virtual (and entirely safe) world.

FACT

English mathematician Alan Turing proposed the "Turing test" for artificial intelligence. This involves a remote interrogator asking questions of a subject who may or may not be human. If, after five minutes of questioning, the interrogator cannot determine whether or not the subject is a computer, then the remote entity is said to be sentient.

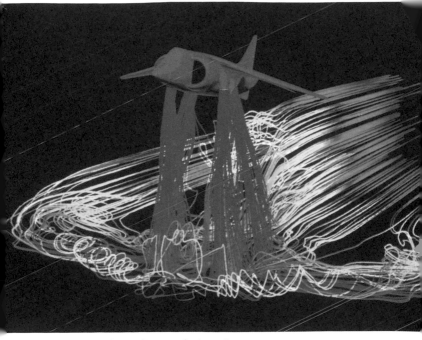

Computer simulation of jet aircraft taking off.

Similarly, the need to produce a scale model of a new aircraft for testing in a costly wind tunnel has been removed by the development of computer modeling. Air flow at altitude over the surface of a wing surface can be measured without so much as stepping outside. This means that much of the truly dangerous work associated with the development of aircraft and other vehicles can take place in a world where the greatest danger is that of the need to reboot.

Even meteorologists have gotten in on the act, using virtual models of the Earth's atmosphere to predict (with uncanny inaccuracy, considering the costs involved) weather patterns several days in advance.

The Internet

One could be forgiven for thinking that the Internet exists so that fans can swap views about their favorite TV programs. But the real strength of the Internet is that it allows all of us to exchange information rapidly with almost anyone, anywhere in the world. So, just what is this Internet?

Internet History

The Internet has been described as a network of networks. A direct descendant of Arpanet, an attempt by the U.S. military in the early 1970s to ensure the integrity of telecommunications in the event of war, the Internet was originally used by universities and other research institutions to link up their giant mainframe computers and share data.

When Arpanet was extended across the Atlantic in 1973 the opportunity arose to bring in academic institutions from all over Europe. The invention, by English scientist Tim Berners-Lee, of the application known as the World Wide Web (www) in the 1980s brought general public access to this network closer still, and by 1991 the first Internet Service Providers (ISPs) were offering relatively cheap online time via the standard telephone system. The timing of this was fortuitous as the original Arpanet was shut down in 1990 – or so the military claim.

How it Works

It might help when imagining the command structure of the Internet to think of a typical family tree. At the top are the **routers**. These are computers that communicate via high-speed fiber-optic cables and satellite links. The routers, although constituting a network in their own right, act as links for the various networks that exist around the world. These networks are controlled largely by ISPs and so play host to millions of home computeres, providing Internet access on every continent in the world.

Data, be it in the form of emails, web pages or attachments, travels around the Internet in "packets." These are relatively small amounts of information that have been given a digital label which describes the destination address of the packets. Once released onto the net, these packets seek out the fastest route to this address where they are then reassembled in the correct order to form a complete file.

Key Terms Associated with the Internet

Term	Definition
TCP/IP	Transmission Control Protocol/Internet Protocol, the system which allows communication between networks
URL	Uniform Resource Locator, the unique address given to a web page
HTML	HyperText Markup Language, the current principal programming language for web pages
HTTP	HyperText Transfer Protocol, the system which allows the movement of hypertext web pages across the Internet
ISP	Internet Service Provider, companies that sell access to the Internet via their network
JPEG	Joint Photographic Experts Group, a standard compressed file format agreed by ISPs for the transmission of digital images across the Internet
MPEG	Moving Pictures Expert Group, video file compression courtesy of Hollywood
MODEM	Modulator-Demodulator, a digital-to-analog (and back again) converter that allows a computer to access the Internet via the telephone system
POP	Point of Presence, the telephone number of your ISP, which your computer calls every time you make a dial-up connection to the Internet
MP3	Compressed music file format which reduces the data for CD tracks to around a thirtieth of their original size

Viruses

Computer viruses, as the name suggests, act much like the viruses that humans and other animals pass between each other. They are essentially self-replicating programs that hinder or otherwise harm the good operation of a computer and/or a computer network.

The viral program often sneaks in on the back of legitimate files, and for a while the most common way of sending a virus was as an email attachment – a kind of modern Trojan Horse. Most users are now aware of this and so tend no longer to open attachments from unknown senders. Once the virus has wormed its way into the computer's operating system it is able to instruct the host program to do any of a number of things. This might be something relatively harmless but annoying or could equally render the computer useless.

A viral infection can be spread via data storage devices, computer networks, or just about any unprotected online system, often with quite devastating results. In 2000 the far from lovely "Love Bug" virus infected the U.S. Congress and the British Parliament, among others, crippling their e-mail systems. Most computer networks and very many home systems now sit behind **firewalls** and so are protected against viruses so long as the protection is updated regularly, and ideally on a daily basis.

Rates of Internet Usage

Almost as soon as access to the Internet was offered commercially to those who had computers at home, its rate of growth all but exploded. From just a few hundred sites in the early 1990s, the net very quickly grew to embrace nearly 25 million websites by the end of the century.

Meanwhile, the number of people with registered Internet addresses rose from just over a million in 1993 to almost 100 million in the same period of time. With well over 300 million users at present, it is no surprise to learn that academics have set up a parallel Internet, known as Abilene, in order to speed up communication between teams of researchers.

LOGIC,
CHAOS THEORY,
& FRACTALS

Plato's Influence on Mathematics

Although known primarily as a philosopher, Plato (c. 400 B.C.) also had a tremendous interest in mathematics. Like others of his time, he believed that the secrets of the Universe were buried in numbers and forms, just waiting to be unearthed.

Plato had an enormous influence on the mathematicians of his day and on those who came after him. So it was that this period saw the emergence of many of the definitions, postulates, and axioms that we associate with traditional mathematics.

Euclidean Mathematics

But when we talk of traditional mathematics we are really talking of Euclid. Now seen as the most important of the Greek mathematicians, Euclid (c. 300 B.C.) was responsible for *The Elements*, a 13-part thesis in which he laid the foundations of geometry by means of a few simple axioms.

So influential was this work that it remained without serious challenge until the nineteenth century and the emergence of a new view of the Universe. Mathematics up to this point is neatly defined as Euclidean mathematics; the new mathematics of the nineteenth century and the subsequent developments that it inspired have quite sensibly been termed non-Euclidean mathematics.

Non-Euclidean Mathematics

The essential weakness in Euclidean mathematics lay in its treatment of two- and three-dimensional figures. To draw a triangle in the sand and then analyze its properties in Euclidean terms is to invite error because under the Euclidean system no account is taken of the fact that the triangle is drawn on a curved surface (our planet is essentially a sphere).

The German mathematician Carl Friedrich Gauss was perhaps the first of the nineteenth-century mathematicians to "doubt the truth of geometry," but once the foundations had been undermined the Euclidean system began to collapse. The final and conclusive push came from Bernhard Riemann, who developed Gauss's ideas on the intrinsic curvature of surfaces.

Riemann argued that we should ignore Euclidean geometry and treat each surface by itself. This had a profound impact on mathematics, removing *a priori* reasoning and ensuring that any future investigation of the geometric nature of the Universe would have to be, at least in part, empirical. It also provided a mechanism for examinations of multidimensional space using an adaptation of the calculus.

Physics and Field Equations

Another fascinating development in mathematics in the nineteenth century was its rigorous application to a new branch of science later to be called physics.

When Michael Faraday was busy discovering electromagnetic induction and rotation in the first half of the nineteenth century he did so in the guise of a natural philosopher. Physics, as a branch of science, had yet to be given a name, and although Faraday is often thought of as one of the first physicists, in truth this remarkable man was nothing of the sort. His mathematical skills, by his own admission, simply weren't up to it.

It took a Scottish mathematician by the name of James Clerk Maxwell to supply a mathematical interpretation of Faraday's work on electromagnetism. Having done that he went on to examine the properties of electromagnetic fields. He was able to conclude mathematically that electromagnetic waves move at the speed of light and that light is just one form of electromagnetic wave.

Maxwell's ideas were both radical and challenging (many of his contemporaries had difficulties with the math). They were also largely ignored, until 1886, when the German physicist Heinrich Hertz confirmed the existence of electromagnetic waves – in this case radio waves – traveling at the speed of light.

Maxwell created four field equations in his examination of the nature of electromagnetism which survived the challenges presented in the twentieth century first by relativity and then by quantum mechanics.

Chaos Theory

If a butterfly flaps its wings in the steamy interior of the Amazon jungle, does that mean that you're going to be losing your roof to a hurricane? Possibly …

Chaos theory is concerned with behavior in chaotic systems, i.e. those systems where very many factors may or may not have an effect on the ultimate outcome. Some systems, such as our planet's weather, are so sensitive to initial inputs (perhaps even the flapping of a butterfly's wings) that it is all but impossible to predict what their outcomes may be. It is for this reason that we cannot really make accurate long-term predictions about the weather.

Other systems which are equally prone to chaotic behavior include those associated with economics. After all, if it were possible to predict accurately the long-term behavior of the stock market you'd be looking at a blank page and I would be viewing properties on the Upper East Side.

The notion of chaos is nothing new. The Taoists believe that the Universe is shaped from primordial chaos, with a continual and unpredictable swing between yin and yang. Early pre-scientific civilizations attributed to the gods the apparently chaotic nature of things.

Science, until recently, was concerned with trying to make sense of this apparent chaos in the hope that the Universe could be made to fit an almost clockwork model where everything could be quantified (and predicted) using a combination of empirical evidence and mathematics. By the twentieth century, however, it was becoming clear that this might not be possible.

Chaos theory recognizes the unpredictable nature of certain systems but embraces chaos as the key to understanding them. It is a new branch of mathematics that may yet provide the key that unlocks the secrets of the Universe.

Strange Attractors and the Butterfly Effect

Most simple graphs of ongoing systems can be adapted to display simple attractors. These are regular loops used to plot predictable, repetitious cycles of behavior. However, the loop begins to stretch and deform if the system becomes chaotic. There is a special kind of graph called a strange attractor that can be used to plot chaotic systems. It does not resemble a normal graph because it is considerably more complex than most, but it does possess a beauty all of its own.

The graph below is the Lorenz Strange Attractor, which is based on possible weather conditions. Things become a little spooky when one notices that the unpredictable yet intricate nature of a weather system produces a graph which bears a remarkable resemblance to a butterfly. This particular strange attractor is named for its creator, meteorologist Edward Lorenz, the man who first suggested that the flapping of a butterfly's wings could have a dramatic influence on the weather.

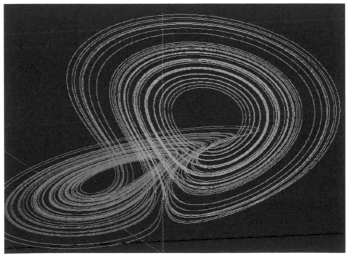

Graph of Lorenz Strange Attractor.

Applied Chaos

Chaos theory is all well and good, but what is it for? What can it do for us? It's still early days, but so far chaos theory has been applied to each of the following.

Applications of Chaos Theory
- turbulence
- irregular heartbeats
- convection patterns (especially those associated with the Sun)
- holes in the asteroid belt
- dripping taps
- weather patterns
- the nervous system
- the motion of gas
- quantum uncertainty

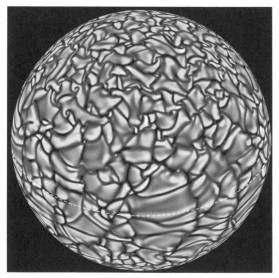

Image of convective patterns from a numerical simulation.

Fractals

As far as mathematicians are concerned, a fractal is a complex geometrical shape. The key concept in the study of fractals is that of self-similarity. A self-similar object is one where each of the component parts resembles the whole. One can see this in nature in, among many other things, a fern frond. If you look closely at the frond you will see that even the smallest part resembles the whole. Astonishingly, this pattern is repeated down to the microscopic level.

Fractal geometry can be observed in the details of a fern frond.

Patterns in fronds were really just sitting there waiting to be discovered. But there are other patterns in nature that proved far more elusive – until now. We've all watched clouds, but who among us can tell how large or small they are? Is that a large cloud floating far away or a small one up close?

The thing is that it makes no difference to the cloud. It can be almost any size from a mile to a hundred miles across and it would still behave like a cloud. This **scale independence** can be plotted statistically, producing something called **statistical self-similarity**. Perhaps the most exciting thing about scale independence is that it can be seen in many of the shapes that nature throws up in an apparently chaotic fashion, such as those of mountains, trees, and maybe even the distribution of matter in the Universe.

Fractal Geometry

Fractals are very useful for describing non-Euclidean, irregular shapes and as such the concept of the fractal has triggered the rise of a new system of geometry. Fractal geometry has since escaped the bounds of mathematics and entered diverse worlds such as physical chemistry and fluid mechanics.

One of the most exciting areas where fractal geometry has been applied is in the field of statistical mechanics. It has proved itself to be especially well-suited to examinations of apparently chaotic systems, such as the distribution of galaxy clusters throughout the Universe.

Fractal geometry is applied to the formations of galaxy clusters.

The Von Koch Curve

The Von Koch Curve is a relatively simple fractal that is created by the **iteration** (repetition) of an equilateral triangle. To begin with, one equilateral triangle is placed upside down over another. The external sides are then divided into three, and new equilateral triangles are constructed using the middle third of each side as the base. This process is then repeated until the figure takes on the shape of a fractal curve.

First four orders of Koch curve.

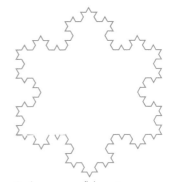

Koch curve snowflake pattern.

Mandelbrot Set

The word "fractal" was coined in 1975 by the Polish-born mathematician Benoit Mandelbrot. (The word was derived from the Latin *fractus*, meaning fragmented.) He devised the most famous of the fractals, named the Mandelbrot Set. This fractal is a beautiful creature that exists only in the rarefied world of pure mathematics, although we can see it with the aid of computer graphics.

The Mandelbrot Set was created by iterating the following equation:

$$z = z^2 + c$$

This equation is deceptively simple in appearance but the variables z and c are actually complex numbers. In the computer image of the Mandelbrot Set on page 118, the iterated equation has produced different final values, each of which is given a different color.

This would not have been possible without the use of a modern computer, but then again the modern computer would not have been able to display the Mandelbrot Set without the aid of fractal algorithms. Fractal algorithms are a development of fractal geometry and have been used to reproduce the complex, irregular shapes of nature – and much more besides – on our computer screens.

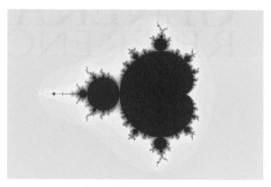

Computer-generated Mandelbrot Set.

If you take a close look at any part of the Mandelbrot Set you will see that it conforms to the rule of self-similarity and closely resembles chaotic forms in nature, such as those associated with cracks in sheets of ice or the frost formations one finds on windows in winter.

Julia Set

Julia Sets (which are named for Gaston Julia) are fractal shapes that are defined on the complex number plane. They are closely related to the Mandelbrot Set and can, in a similar fashion, be plotted with different colors to indicate the various iterations that have taken place.

Fractal Dimension

A key characteristic of the fractal – one that is vital to our understanding of complex non-Euclidean figures – is that of fractal dimension. Dimension in the world of fractals differs from that of the Euclidean notion of dimension. Fractal dimension is a parameter that remains the same no matter how much an object is magnified or where it is viewed from. If we look at a fractal curve we can see that at each stage in its construction the perimeter of the curve increases in the ratio 4 to 3. If we call the fractal dimension D, then we can see that this figure (D) is the amount that the perimeter must increase in order to be raised from 3 to 4, producing the equation:

$$3^D = 4$$

GENERAL
REFERENCE

Squares and Cubes

Numeral	Square	Cube	Numeral	Square	Cube
1	1	1	11	121	1,331
2	4	8	12	144	1,728
3	9	27	13	169	2,197
4	16	64	14	196	2,744
5	25	125	15	225	3,375
6	36	216	16	256	4,096
7	49	343	17	289	4,913
8	64	512	18	324	5,832
9	81	729	19	361	6,859
10	100	1,000	20	400	8,000

Roman Numerals

Roman	Arabic	Roman	Arabic	Roman	Arabic
I	1	XV	15	XC	90
II	2	XIX	19	IC	99
III	3	XX	20	C	100
IV	4	XXIX	29	CIC	199
V	5	XXX	30	CC	200
VI	6	XL	40	CD	400
VII	7	IL	49	D	500
VIII	8	L	50	DC	600
IX	9	LIX	59	DCC	700
X	10	LX	60	CMLIII	953
XI	11	LXVIII	68	M	1,000
XIV	14	LXIX	69	MMIX	2,009

Areas and Volumes of Two- and Three-dimensional Shapes

Two-dimensional Shapes

Circle
r = radius

circumference = 2πr
area = πr²

Triangle
h = height
a, b, c = sides
b = base

perimeter = a + b + c
area = ¹/₂bh

Rectangle
a, b = sides

perimeter = 2(a + b)
area = ab

Three-dimensional Shapes

Cylinder
h = height
r = radius

surface area =
2πr² + 2πrh
volume = πr²h

Cone
h = height
r = radius
l = side

surface area = πrl + πr²
volume = ¹/₃πr²h

Cube
a = side

surface area = 6a²
volume = a³

Rectangular Prism
a, b, c = sides

surface area =
2(ab + bc + ac)
volume = abc

Sphere
r = radius

surface area = 4πr²
volume = ⁴/₃πr³

Square Pyramid
a = side along base
h = height

surface area = 6a²
volume = ¹/₃a²h

...ersion Tables

...tric to Imperial Conversions

To convert	Into	Multiply by
centimeters	inches	0.3937
meters	feet	3.281
kilometers	miles	0.6214
meters	yards	1.094
grams	ounces	0.03527
kilograms	pounds	2.205
tonnes	tons	0.9843
square centimeters	square inches	0.155
square meters	square feet	10.76
hectares	acres	2.471
square kilometers	square miles	0.3861
square meters	square yards	1.196
cubic centimeters	cubic inches	0.06102
cubic meters	cubic feet	35.31
liters	pints	1.76
liters	gallons	0.22

Weights for Cooking

Imperial	Metric	Imperial	Metric	Imperial	Metric
$\frac{1}{4}$ oz	10 g	1$\frac{3}{4}$ oz	50 g	2 lb	900 g
$\frac{1}{2}$ oz	15 g	2 oz	55 g	2 lb 4 oz	1 kg
$\frac{3}{4}$ oz	20 g	3 oz	85 g	2 lb 12 oz	1.25 kg
1 oz	25 g	4 oz	115 g	3 lb 5 oz	1.5 kg
1$\frac{1}{4}$ oz	35 g	5 oz	140 g	4 lb 8 oz	2 kg
1$\frac{1}{2}$ oz	40 g	1 lb	450 g	6 lb 8 oz	3 kg

Imperial to Metric Conversions

To convert	Into	Multiply by
inches	centimeters	2.54
feet	meters	0.3048
miles	kilometers	1.609
yards	meters	0.9144
ounces	grams	28.35
pounds	kilograms	0.4536
tons	tonnes	1.016
square inches	square centimeters	6.452
square feet	square meters	0.0929
acres	hectare	0.4047
square miles	square kilometers	2.59
square yards	square meters	0.8361
cubic inches	cubic centimeters	16.39
cubic feet	cubic meters	0.02832
pints	liters	0.5683
gallons	liters	4.546

Dry Volumes for Cooking

Imperial	Metric	Imperial	Metric
1 tsp	5 ml	½ cup	125 ml
3 tsp/1 tbsp	15 ml	⅔ cup	150 ml
2 tbsp	30 ml	¾ cup	175 ml
3 tbsp	45 ml	1 cup	250 ml
¼ cup	60 ml	2 cups	500 ml
⅓ cup	70 ml	4 cups	1 l

Liquid Volumes for Cooking

Imperial	Metric	Cups
1 fl oz	25 ml	
2 fl oz	50 ml	¼ cup
2½ fl oz	75 ml	
4 fl oz	125 ml	½ cup
5 fl oz/¼ pint	150 ml	
6 fl oz	175 ml	¾ cup
10 fl oz/½ pint	300 ml	1¼ cups
12 fl oz	350 ml	1½ cups
20 fl oz/1 pint	568 ml	2½ cups
1½ pints	850 ml	3¾ cups

SI Units and Definitions

SI units	Definition
1 meter	The distance traveled by light in a vacuum in one 299,792,458ths of a second
1 kilogram	A cylinder of platinum-iridium alloy held by the International Bureau of Weights and Measures at Sevres, near Paris. (This is the only remaining artifact-based standard measure still in use.)
1 second	9,192,631,770 radiation cycles of the cesium-133 atom
1 ampere	The magnitude of a current that results in a force equal to 2×10^{-7} newtons per meter of length
1 kelvin	The point immediately above absolute zero, where all atomic activity ceases (the fraction 1/273.16 of the thermodynamic temperature of the triple point of water)
1 candela	The luminous intensity of a source that emits monochromatic radiation of frequency 540×10^{12} Hz
1 mole	The amount of a substance that contains as many elementary entities as there are atoms in 12 grams of carbon-12

SI Measurements

Multiple	Prefix	Symbol
1/1,000,000,000,000,000,000 (10^{-18})	N/A	a
1/1,000,000,000,000,000 (10^{-15})	N/A	f
1/1,000,000,000,000 (10^{-12})	N/A	p
1/1,000,000,000 (10^{-9})	nano	n
1/1,000,000 (10^{-6})	micro	m
1/1,000 (10^{-3})	milli	m
1/100 (10^{-2})	centi	c
1/10 (10^{-1})	deci	d
10	deca	da
100 (10^{2})	hecto	h
1,000 (10^{3})	kilo	k
1,000,000 (10^{6})	mega	M
1,000,000,000 (10^{9})	giga	G
1,000,000,000,000 (10^{12})	tera	T
1,000,000,000,000,000 (10^{15})	peta	P
1,000,000,000,000,000,000 (10^{18})	exa	E

FACT

On 20 May 1875, delegates of 17 countries signed the *Meter Convention*, later renamed *Systeme International d'Unites* (International System of Units) with the aim to implement a standardized system of measures. It was amended in 1921 and again in 1960. Today there are 48 countries that have signed the convention.

SI Quantities

Unit	SI name	Symbol
Absorbed dose of radiation	gray	Gy
Amount of a substance	mole	mol
Electric capacitance	farad	F
Electric charge	coulomb	C
Electric conductance	siemens	S
Electric current	ampere	A*
Electrical resistance	ohm	Ω
Energy or work	joule	J
Force	newton	N
Frequency	hertz	Hz
Illumination	lux	lx
Inductance	henry	H
Luminous flux	lumen	lm
Luminous intensity	candela	cd
Magnetic flux	weber	Wb
Magnetic flux density	tesla	T
Plane angle	radian	rad
Pressure	pascal	Pa
Radiation dose equivalent	sievert	Sv
Radiation exposure	roentgen	r
Radioactivity	becequerel	Bq
Thermodynamic temperature	kelvin	K

* *Guinness Book of Facts* defines this as magnemotive force.

Logarithm Table for Sine, Cosine, and Tangent

Angle	Sin (angle)	Cos (angle)	Tan (angle)
0	0	1	0
1	0.1740	0.9998	0.175
2	0.3490	0.9994	0.0349
3	0.5230	0.9986	0.0524
4	0.6980	0.9976	0.0699
5	0.8720	0.9962	0.0875
6	0.1045	0.9945	0.1051
7	0.1219	0.9926	0.1228
8	0.1392	0.9903	0.1405
9	0.1564	0.9877	0.1584
10	0.1736	0.9848	0.1763
11	0.1908	0.9816	0.1944
12	0.2079	0.9781	0.2126
13	0.2249	0.9744	0.2309
14	0.2419	0.9703	0.2493
15	0.2588	0.9659	0.2679
16	0.2756	0.9613	0.2867
17	0.2924	0.9563	0.3057
18	0.3090	0.9511	0.3249
19	0.3256	0.9455	0.3443
20	0.3420	0.9397	0.3640
21	0.3584	0.9336	0.3839
22	0.3746	0.9272	0.4040
23	0.3907	0.9205	0.4245
24	0.4067	0.9135	0.4452
25	0.4226	0.9063	0.4663
26	0.4384	0.8988	0.4877
27	0.4540	0.8910	0.5095
28	0.4695	0.8829	0.5317
29	0.4848	0.8746	0.5543
30	0.5000	0.8660	0.5773

Logarithm Table for Sine, Cosine, and Tangent (cont'd)

Angle	Sin (angle)	Cos (angle)	Tan (angle)
31	0.5150	0.8571	0.6009
32	0.5299	0.8480	0.6249
33	0.5446	0.8387	0.6494
34	0.5592	0.8290	0.6745
35	0.5736	0.8191	0.7002
36	0.5878	0.8090	0.7265
37	0.6018	0.7986	0.7535
38	0.6157	0.7880	0.7813
39	0.6293	0.7772	0.8098
40	0.6428	0.7660	0.8391
41	0.6561	0.7547	0.8693
42	0.6691	0.7431	0.9004
43	0.6820	0.7314	0.9325
44	0.6947	0.7193	0.9657
45	0.7071	0.7071	1.0000
46	0.7193	0.6947	1.0355
47	0.7314	0.6820	1.0724
48	0.7431	0.6691	1.1106
49	0.7547	0.6561	1.1504
50	0.7660	0.6428	1.1918
51	0.7772	0.6293	1.2349
52	0.7880	0.6157	1.2799
53	0.7986	0.6018	1.3270
54	0.8090	0.5878	1.3764
55	0.8191	0.5736	1.4281
56	0.8290	0.5592	1.4826
57	0.8387	0.5446	1.5399
58	0.8480	0.5299	1.6003
59	0.8571	0.5150	1.6643
60	0.8660	0.5000	1.7321

Angle	Sin (angle)	Cos (angle)	Tan (angle)
61	0.8746	0.4848	1.8040
62	0.8829	0.4695	1.8907
63	0.8910	0.4540	1.9626
64	0.8988	0.4384	2.0503
65	0.9063	0.4226	2.1445
66	0.9135	0.4067	2.2460
67	0.9205	0.3907	2.3559
68	0.9272	0.3746	2.4751
69	0.9336	0.3584	2.6051
70	0.9397	0.3420	2.7475
71	0.9455	0.3256	2.9042
72	0.9511	0.3090	3.0777
73	0.9563	0.2924	3.2709
74	0.9613	0.2756	3.4874
75	0.9659	0.2588	3.7321
76	0.9703	0.2419	4.0108
77	0.9744	0.2249	4.3315
78	0.9781	0.2079	4.7046
79	0.9816	0.1908	5.1446
80	0.9848	0.1736	5.6713
81	0.9877	0.1564	6.3138
82	0.9903	0.1392	7.1154
83	0.9926	0.1219	8.1443
84	0.9945	0.1045	9.5144
85	0.9962	0.0872	11.4300
86	0.9976	0.0698	14.3010
87	0.9986	0.0523	19.0810
88	0.9994	0.0349	28.6360
89	0.9998	0.0174	57.2900
90	1	0	infinite

Mathematical Symbols

Symbol	Meaning	Symbol	Meaning
+	Addition	∞	Infinity
−	Subtraction	Σ	Summation
×	Multiplication	v, $\underline{v}$	Vectors
÷	Division	f(×)	Function
=	Equals	!	Factorial
≠	Does not equal	√	Square root
>	Greater than	A∩B	Intersection
<	Less than	A∪B	Union
≥	Greater than or equal to	A⊂B	Subset
≤	Less than or equal to	∅	Null set

Basic Rules of Algebra

Expression	Action	Expression becomes
$a + a$	Simple addition	$2a$
$a + b = c + d$	Subtract b from either side	$a = c + d - b$
$ab = cd$	Divide both sides by b	$a = cd/b$
$(a + b)(c + d)$	Multiplication of bracketed terms	$ac + ad + bc + bd$
$a^2 + ab$	Use parentheses	$a(a + b)$
$(a + b)^2$	Expand brackets	$a^2 + 2ab + b^2$
$a^2 - b^2$	Difference of two squares	$(a + b)(a - b)$
$1/a + 1/b$	Find common denominator	$(a + b)/ab$
$a/b \div c/d$	Dividing by a fraction is the same as multiplying by its reciprocal	$a/b \times d/c$

Elements in the Earth's Crust

Element	Mass (%)	Element	Mass (%)
Oxygen	49.13	Sodium	2.40
Silicon	26.00	Potassium	2.35
Aliminum	7.45	Magnesium	2.35
Iron	4.20	Hydrogen	1.00
Calcium	3.25	Others	1.87

Common Names and Formulae of Important Compounds

Common name	Chemical name	Formula
Water	Hydrogen oxide	H_2O
Salt	Sodium chloride	$NaCl$
Bicarbonate of soda	Sodium hydrogencarbonate	$NaHCO_3$
Household bleach	Sodium chlorate (I)	$NaOCl$
Methylated spirits	Methanol	CH_3OH
Alcohol	Ethanol	C_2H_5OH
Vinegar	Ethanoic acid	$CH_3.COOH$
Vitamin C	Ascorbic acid	$C_4H_5O_4.CHOH.CH_2OH$
Aspirin	Acetylsalicylic acid	$C_6H_4.COOCH_3.COOH$
White sugar	Sucrose	$C_6H_{11}O_5.O.C_6H_{11}O_5$
Limestone/chalk	Calcium carbonate	$CaCO_3$
Rust	Hydrated iron (III) oxide	$Fe_2O_3.xH_2O$

Temperature Scales

To convert	Into	Equation
Celsius (C)	Fahrenheit (F)	$F = (C \times 9 \div 5) + 32$
Fahrenheit	Celsius	$C = (F - 32) \times 5 \div 9$
Celsius	Kelvin (K)	$K = C + 273$
Kelvin	Celsius	$C = K - 273$
Fahrenheit	Kelvin	$K = ((F - 32) \times 5 \div 9) + 273$

Melting and Boiling Points of Elements

Element	Melting point °C	°F	Boiling point °C	°F
Mercury	−39	−38	357	675
Helium	−272	−458	−269	−452
Tungsten	3,410	6,170	5,555	10,031
Nitrogen	−210	−346	−196	−321
Sodium	98	208	883	1,621
Oxygen	−219	−362	−183	−297
Bromine	−7	19	59	138
Iron	1,535	2,795	2,862	5,184
Carbon	3,550	6,420	4,827	8,720
Gold	1,063	1,945	2,970	5,379

Recent Discovery of Elements

Element name	Pure element first isolated	Origin of name
Cobalt, Co	1735 by Georg Brandt	German *kobold*, meaning goblin
Hydrogen, H	1766 by Henry Cavendish	Greek *hydro-* and *genes*, meaning water-maker
Chlorine, Cl	1774 by Karl Wilhelm Scheele	Greek *chloros*, meaning greenish-yellow
Tungsten, W	1783 by Juan José and Fausto Elhuyar	Swedish *tung*, meaning heavy, and *sten*, meaning stone
Chromium, Cr	1797 by Nicolas-Louis Vauquelin	Greek *chroma*, meaning colour
Bromine, Br	1826 by Antoine-Jérôme Balard	Greek *bromos*, meaning stench
Helium, He	1868 by Pierre Janssen and Norman Lockyer	Greek *helios*, meaning the Sun
Rutherfordium, Rf	1964 (in U.S.S.R.) and 1969 (in U.S.A.)	Named after New Zealander Ernest Rutherford

Ions and Radicals

Name	Formula and charge	Name	Formula and charge
Hydrogen	H^+	Silver (I)	Ag^+
Sodium	Na^+	Zinc	Zn^{2+}
Potassium	K^+	Ammonium	NH_4^+
Magnesium	Mg^{2+}	Hydroxonium	H_3O^+
Calcium	Ca^{2+}	Oxide	O^{2-}
Aluminum	Al^{3+}	Sulfide	S^{2-}
Iron (II)	Fe^{2+}	Fluoride	F^-
Iron (III)	Fe^{3+}	Chloride	Cl^-
Copper (I)	Cu^+	Iodide	I^-
Copper (II)	Cu^{2+}	Hydroxide	OH^-

Local Group of Galaxies

Name	Type	Distance (light years)	Luminosity (million Suns)	Diameter (light years)
Milky Way	Spiral	0	15,000	100,000
Large Magellanic Cloud	Irregular Spiral	170,000	2,000	30,000
Small Magellanic Cloud	Irregular	190,000	500	20,000
Sculptor	Elliptical	300,000	1	6,000
Carina	Elliptical	300,000	0.01	3,000
Draco	Elliptical	300,000	0.1	3,000
Sextans	Elliptical	300,000	0.01	3,000
Ursa Minor	Elliptical	300,000	0.1	2,000
Fornax	Elliptical	500,000	12	6,000
Leo I	Elliptical	600,000	0.6	2,000
Leo II	Elliptical	600,000	0.4	2,000
NGC 6822	Irregular	1,800,000	90	15,000
IC 5152	Irregular	2,000,000	60	3,000
WLM	Irregular	2,000,000	90	6,000
Andromeda (M 31)	Spiral	2,200,000	40,000	150,000
Andromeda I, II, III	Elliptical	2,200,000	1	5,000
M 32 (NGC 221)	Elliptical	2,200,000	130	5,000
NGC 147	Elliptical	2,200,000	50	8,000
NGC 185	Elliptical	2,200,000	60	8,000
NGC 205	Elliptical	2,200,000	160	11,000
M 33 (Triangulum)	Spiral	2,400,000	5,000	40,000
IC 1613	Irregular	2,500,000	50	10,000
DDO 210	Irregular	3,000,000	2	5,000
Pisces	Irregular	3,000,000	0.6	2,000
GR 8	Irregular	4,000,000	2	1,500
IC 10	Irregular	4,000,000	250	6,000
Sagittarius	Irregular	4,000,000	1	4,000
Leo A	Irregular	5,000,000	20	7,000
Pegasus	Irregular	5,000,000	20	7,000

Brightest Stars

Name	Constellation magnitude	Apparent magnitude	Absolute magnitude	Distance (light years)	Star type
Sun	N/A	−26.7	4.8	0.000015	Yellow main sequence
Sirius A	Canis Major (The Great Dog)	−1.4	1.4	8.6	White main sequence
Canopus	Carina (The Keel)	−0.7	−8.5	1,200	White supergiant
Alpha Centauri A	Centaurus (The Centaur)	−0.1	4.1	4.3	Yellow main sequence
Arcturus	Boötes (The Herdsman)	−0.1	−0.3	37	Red giant
Vega	Lyra (The Lyre)	0.04	0.5	27	White main sequence
Capella	Auriga (The Charioteer)	0.1	−0.6	45	Yellow giant
Rigel	Orion (The Huntsman)	0.1	−7.1	540–900	White supergiant
Procyon	Canis Minor (The Little Dog)	0.4	2.7	11.3	Yello main sequence
Achenar	Eridanus (River Eridanus)	0.5	−1.3	85	White main sequence

Earthquake Measurement

Mercalli scale	Effects
1	Not felt by people but recorded by instruments; doors may swing
2–4	Felt by people indoors and some people outdoors
5–6	Felt by most or all outdoors; buildings tremble, books fall off shelves
7–8	Branches fall off trees and it's difficult to drive
9–10	Cracks appear in roads; buildings and bridges collapse
11–12	Few buildings standing; waves are visible in the ground; rivers may change course

Richter scale	Effects
1–3	Detectable only with instruments
4	Detectable within 20 miles (32 km) of epicenter
5	May cause slight damage to buildings; poorly designed buildings will be damaged
6	Moderately destructive to well-designed buildings
7	A major earthquake
8–9	A very destructive earthquake, causing serious damage up to several hundred miles across

Physics Symbols

Symbol	Meaning	Symbol	Meaning
α	Alpha particle	μ	Micro-; permeability
β	Beta particle	ν	Frequency; neutrino
γ	Gamma ray	ρ	Density; resistivity
ε	Electromotive force	σ	Conductivity
η	Efficiency; viscosity	c	Speed of light
λ	Wavelength		

Physics Formulae

Weight
Weight is equal to mass multiplied by acceleration due to gravity.

$$W = mg$$

Where:
W = weight
m = mass
g = acceleration due to gravity

Pressure
Pressure is equal to force applied divided by area over which force acts.

$$P = \frac{F}{A}$$

Where:
P = pressure
F = applied force
A = area over which force acts

Turning Force
Turning force is equal to force multiplied by distance of applied force from pivot.

$$T = Fd$$

Where:
T = turning force (moment)
F = applied force
d = distance

Newton's Second Law
Acceleration is equal to force divided by mass.

$$a = \frac{F}{m}$$

Where:
a = acceleration
F = applied force
m = mass

Speed
Speed is equal to distance divided by time.

$$v = \frac{d}{t}$$

Where:
v = speed
d = distance
t = time

Constant Acceleration
Acceleration is equal to change in speed divided by time taken for that change.

$$a = \frac{(v_2 - v_1)}{t}$$

Where:
a = acceleration
v_1 = speed at the beginning of the time interval
v_2 = speed at the end of the time interval
t = time

Momentum

Momentum is equal to mass multiplied by speed.

$$p = mv$$

Where:
p = momentum
m = mass
v = speed

Friction

Frictional force between two surfaces is equal to the coefficient of friction multiplied by the force acting to keep the surfaces together.

$$F = \mu N$$

Where:
F = frictional force
μ = coefficient of friction; this varies with different materials
N = forces between two surfaces

Liquid Pressure

Pressure is equal to the liquid's density multiplied by acceleration due to gravity multiplied by height of water above point.

$$P = \rho g h$$

Where:
P = pressure
ρ = liquid pressure
g = acceleration due to gravity
h = height of liquid above measured point

Gravitation

Gravitational force equals a constant multiplied by mass one, multiplied by mass two, divided by the distance between the masses squared.

$$F = \frac{Gm_1 m_2}{d^2}$$

Where:
F = gravitational force between two objects
G = gravitational constant
m_1 = mass of object one
m_2 = mass of object two
d = distance between the two objects

Centripetal Force

Force is equal to mass multiplied by the speed squared divided by the radius

$$F = \frac{mv^2}{r}$$

Where:
F = centripedal force
m = mass of object
v = speed of circular motion
r = radius of object's path

Work

Work is equal to force multiplied by distance.

$$W = Fd$$

Where:
W = work done
F = applied force
d = distance moved in line with force

Elasticity

The extension of a solid is proportional to the force applied to it.

$$F \alpha x$$

Where:
F = applied force
x = extension of solid

Current

Current is equal to voltage divided by resistance.

$$I = \frac{V}{R}$$

Where:
I = current
V = voltage
R = resistance

Power

Power is equal to voltage multiplied by current.

$$P = VI$$

Where:
P = power
V = voltage
I = current

Useful Fractions, Decimals, and Percentages

Fraction	Decimal	Percentage
$1/2$	0.5	50%
$1/4$	0.25	25%
$3/4$	0.75	75%
$1/5$	0.2	20%
$1/10$	0.1	10%
$1/100$	0.01	1%
$1/8$	0.125	$12 1/2$%
$1/3$	0.333 recurring	$33 1/3$%
$2/3$	0.66 recurring	$66 2/3$%

Metric Prefixes

Prefix	Value	Prefix	Value
exa-	10^{18}	deci-	$1/10$
peta-	10^{15}	centi-	$1/100$
tera-	10^{12}	milli-	10^{-3}
giga-	10^{9}	micro-	10^{-6}
mega-	10^{16}	nano-	10^{-9}
kilo-	1,000	pico-	10^{-12}
hecto-	100	femto-	10^{-15}
deca-	10	atto-	10^{-18}

Computer Coding

Decimal number	Binary	Hexadecimal	ASCII character
0	00000000	00	NUL*
1	00000001	01	SOH*
2	00000010	02	Start of text*
9	00001001	09	HT*
10	00001010	0A	Line feed*
11	00001011	0B	VT*
12	00001100	0C	Form feed*
13	00001101	0D	Carriage return*
14	00001110	0E	SO*
15	00001111	0F	SI*
16	00010000	10	DLE*
17	00010001	11	DC1*
18	00010010	12	DC2*
32	00100000	20	SPACE
33	00100001	21	!
34	00100010	22	"
64	01000000	40	@
65	01000001	41	A
66	01000010	42	B
119	01110111	77	w
120	01111000	78	x
121	01111001	79	y
122	01111010	7A	z
123	01111011	7B	{

* Control characters (non-printing)

Index

Page numbers in *italics* refer
to illustrations

3D graphs 43

A

abacus 9, 10, 94
Abilene 108
absolute magnitude 40
acceleration 56
constant 136
al-Khwarizmi, Muhammad
ibn Musa 11, 48
algebra 48–51
basic rules of 130
algorithms, fractal 117
alpha particles 83
amps 73
Andromeda Nebula 88
angles 57–58
annulus 59
apparent magnitude 40
Apple Macintosh 98
arc of circle 59
Archimedes 15, 23, 55
Archimedes's spiral 26
Aristotle 88
Arpanet 106
artificial intelligence 104
ASCII 100
astronomical measurement
38–39
atomic clock 35
atomic mass 39
atomic number 39, 78–79
atoms 35, 75–76, 77, 86–87
averages 44–45

Babbage, Charles: analytical
engine 95
Babylonia: counting system 9
measuring system 30
background radiation 89, 90
bar chart 43
base 2 *see* binary system
base 10 9, 10
base 60 9
BASIC 97

Berners-Lee, Tim 106
beta particles 83
Big Bang Theory 89, 90,
91, 92

B

binary system 9, 100, 101
binomials 49
bitmap 101
Boolean logic 95–97
Boyle's Law 71
Brane Theory 91, 92

C

calculating machines
94–97
calculus 55–57
differential 56
integral 56–57
centripetal force 137
chain reaction 84
chaos theory 112–114
chemistry 75–87
Charles's Law 71
circle 15, 21, 22, 59
area of 15, 55, 121
circumference of 15, 22,
121
clocks 34–35
COBE (Cosmic Background
Explorer) probe 89
coding, *see* computers
complex numbers 15
compounds 77
common names and
formulae of 131
computers 9, 98–100
codes 100, 139
components of 98–99
early 97
and fractals 117
and science 104–105
speed of improvement
103
viruses 107–108
cone
surface area of 121
volume of 121

conic sections 21
constants 48
conversion tables 122–124
coordinate systems
51–52, 54
Copernicus, Nicolas 88
cosine 52, 53, 61
logarithm table of
127–129
cosmology 87–92
cube 22
squares and cubes 120
surface area of 121
volume of 121
current *see* electric current
curves 21
cylinder
surface area of 121
volume of 121

D

dark matter 92
data storage 102–103
decibel scale 40
decimals 12, 16
decimation 17
denary system *see* base 10
Descartes, René 26, 51
diameter of circle 22, *58*
differential calculus 56
digitization 100–103
dimension 19
fractal 118
dimensions 91
dodecahedron 22

E

Earth, mass of 18
earthquakes, measurement
of 135
Egypt
ancient: counting system 9
measuring system 30, 31
Eiffel Tower, to measure
height of 62
Einstein, Albert 80, 82, 92
General Theory of
Relativity 81

Special Theory of
    Relativity 80–81, 84
elasticity, Hooke's law of 67
electric circuits 74
electric current 72–73
    physics formula for 137
electric motor 75
electricity 72–75
    generation of 74–75
    measurement of 73
electromagnetic field 74–75
electromagnetic spectrum
    41, 75
electromagnetism 72, 111
electrons 72–73, 75, 76,
    83, 86
elements 75, 78–79
    in the Earth's crust 131
    melting and boiling
        points of 132
    recent discovery of 133
ellipse 21
email 106, 108
energy 68, 82
    nuclear 84–85
    and relativity 80
equations 48, 49
    field 111–112
equilateral triangle 60
Eratosthenes 88
Eratosthenes's Sieve 14
Euclid: Elements 39, 110
Euclidean mathematics 110
Eudoxus 39
exchequer 32

F
factors 14
Faraday, Michael 74–75,
    111
Fermat's last theorem 51
ferromagnetic materials 74
Fibonacci series 27, 28
figurate numbers 13
fluid dynamics 72
force 66–67
    centripetal force, physics
        formula for 137
    turning force, physics
        formula for 137

four-sided shapes 19
fractals 115–118
fractions 12, 16
French Revolution 33
friction, physics formula
    for 137
functions 52, 56
    graphs of 52–54
    polar 54
fundamental interval 69

G
galaxies
    and fractal geometry 116
    local group of 134
    spiral 27, 27
Galileo Galilei 88
gamma rays 83, 84
Gamow, George 89
gas laws 71
gases 70, 71–72
Gauss, Carl Friedrich
    110–111
Gaussian curve 43, 45
General Theory of
    Relativity 81
Geometry 19, 110–111
    fractal 116
    see also circle;
        quadrilaterals; triangle
Golden Mean 26
graphical user interface 98
graphs 41–43, 45
    of the functions 52–54
gravitation, physics formula
    for 137
gravity
    Brane Theory of 91
    General Theory of
        Relativity 81
    Newton's law of
        66–67, 81
Greenwich Mean Time
    36–37

H
heat 68
    measurement of see
        temperature scales
heptagon 20

Hertz, Heinrich 111
hexagon 20
Hipparchus 60
Hooke's Law 67
Hoyle, Fred 90
Hubble, Edwin 88, 89
Hubble's Constant 89
Hubble's Law 88–89
Hypatia 21
hyperbola 21

I
icosahedron 22
imaginary numbers 15
imperial system see
    measuring systems
indices 17
infinitesimals 11
infinity 10
integers 12
integral calculus 56–57
integrated circuit chip 98
Internet 106–108
inverse square law 66–67
ions 76
irrational numbers 12, 15
isosceles triangle 60
iteration 116, 117, 118

J
Jacquard, Joseph-Marie:
    automated loom 95
Julia Set 118

K
kelvins 70
Kepler, Johannes 94
    laws of planetary motion
        65–66
Königsberg Bridge
    problem 24

L
Leibniz, Gottfried 55
    calculating machine 95
Lemaître, Georges 89
Leonardo da Vinci 26
light, speed of 38, 82, 111
limit, concept of 55
line graph 42

liquids 70, 72
and pressure 72, 137
logarithms 17
table of for sine, cosine,
and tangent 127–129
Lorenz, Edward: strange
attractor 113

M

magnetism 74
see also electromagnetism
magnitude, apparent and
absolute 40
Mandelbrot Set 117–118,
118
mass: atomic 39
of Earth and planets 18
and gravitation 81
molecular 40
and relativity 80–81
mathematical symbols 130
Maxwell, James Clerk 111
mean 44
measurement
astronomical 38–39
of earthquakes 135
of electricity 73
of pressure 70
scientific 39–41, 124–126
of temperature 69–70
theory of 39
of time 34–37
see also measuring systems
measuring systems
Babylonian 30–31
British 32
Egyptian 31
imperial 122–124
medieval 32
metric 33, 122–124
Roman 31, 32
SI units 33
median 45
membranes *see* Brane Theory
Mercalli scale 135
metric system *see* measuring
systems
Microsoft 98
Milky Way 27, 88, 92
Möbius strip 25

mode 44
molecular mass 41
molecules 76–77
momentum, physics formula
for 137
monomials 48
Moon, phases of 34, *34–35*
Moore's Law 103
motion
Newton's laws of 64–65
planetary 65–66

N

natural numbers 12
neutrons 83, 84, 87
Newton, Sir Isaac 55, 64,
66, 81
Newton's Laws 64–67
formula for 136
newtons 70
non-Euclidean mathematics
110–111
notation, scientific 18
nuclear bomb 82, 84
nuclear fission 84, *85*
nuclear fusion 84–85
number systems 10–11
numbers: complex 15
imaginary 15
irrational 12, 15
natural 12
prime 14
rational 12
square 13
triangular 13
see also integers
numerals, Roman 120

O

octagon *20*
octahedron 22
Ohm's Law 73
Olbers's Paradox 92
orbitals 86

P

Pacioli, Lucas 26
parabola 21
parallel circuit 74
parallelogram *19*

particle physics 86–87
Pascal, Blaise: calculating
machine 94–95
pendulum 35
pentagon *20*
Penzias, Arno 89
percentages 16–17, 138
Periodic Table 77, 78–79
physics 64–92
formulae 136–137
symbols of 136
pi (π) 12, 15, 22
pictures, digital 101–102
pie chart 41
pixels 101–102
planets: mass of 18
motion of 65–66
Plato 110
polar coordinates 54
polygons 20, 55
polyhedrons 22–23
polynomials 48
power, physics formula
for 137
powers 17, 18
prefixes, metric 138
pressure 70–72
physics formula for 136
physics formula for liquid
pressure 137
Pressure Law 71
prime numbers 14
probability 44–46
protons 83, 86
pyramid 23, *23*
surface area of 121
volume of 121
Pythagorean Theorem 49,
50, 52

Q

quadratic equations 49
quadrilaterals 19
quantum loops 87
quarks 87
quartz technology 35

R

radiation 83
background 89, 90

radicals 133
radioactive decay 83
radius of circle 22, *58*
rational numbers 12
rectangle *19*, 55
  area of 121
  perimeter of 121
rectangular prism *121*
  surface area of 121
  volume of 121
red shift 89
re-entrant polygons 20
relativity 80–81, 112
resistance 73
rhombus *19*
Richter scale 135
Riemann, Bernhard 111
right-angled triangle 50,
  60, 61, 62
Roman numerals 120
Rome, ancient: measuring
  system 31, 32

S
sampling, digital 101
scale independence 115
scalene triangle *60*
scatter diagram 42
Schickard, Wilhelm 94
scientific measurement
  39–41
scientific notation 18
sector of circle *59*
segment of circle *59*
self-similarity *see* fractals
semicircle *59*
series circuit 74
shapes
  three-dimensional 121
  two-dimensional 19–22,
    121
  see also circle;
    quadrilaterals; triangle
SI
  measurements 125
  quantities 126
  units and definitions 33,
    124
simultaneous equations 49
sine 52, 53, 61

logarithm table of
  127–129
Solar System *38*, 39
solid figures 22–23
solids 70
sound
  digital 100–101
  measurement of 40
space–time 80, 81
Special Theory of Relativity
  80–81, 84
speed, physics formula for
  136
sphere 22–23
  surface area of 121
  volume of 121
spirals 26, 27
square *19*, 20
  numbers 13
  pyramid 23, *23*, 121
  roots 14
stars, list of brightest 135
statistics 44–46
Steady State Theory 90
strange attractors 113, *113*
string theory 87
subatomic particles 83,
  86–87
sundials 34
superstrings 87
symbols
  mathematical 130
  physics 136
*Systeme International
  d'Unites* 125

T
tally sticks 8
tangent 52, 54, 61
  of circle *58*
  logarithm table of
    127–129
temperature scales 68–70,
  132
terms 48–49
tetrahedron 22, 23
time, measurement of
  34–37
time zones *36–37*
topology 24–25, 43

trapezium *19*
triangle 19, 20, 60–62, 121
  area of 121
  perimeter of 121
  right-angled 50, *60*,
    61, 62
  types of 60
triangular numbers 13
trigonometry 60–62
trinomials 48
Turing test 104
Tuvalu: domain suffix 99

U
Universal Time 37
Universe: chaos theory
  112–114
  expanding 89
  fundamental forces of 91
  to measure 38–39
  nature of 88, 91–92
  size of 88–89, 92
  theories about origin of
    89–91
upthrust 72

V
variables 48
vector coordinates 51–52
velocity 56
video, digital 102
virtual modeling 104–105
volts/voltage 73
volume 23, 55
volumes
  dry measurements for
    cooking 123
  liquid measurements for
    cooking 124
Von Koch curve 116, *117*

W
weight, physics formula
  for 136
Wilson, Robert 89
work, physics formula for
  137

Z
zero 10, 11

Picture Credits

The publishers would like to thank the following for permission to reproduce images.

Global Book Publishing Photo Library: Close-up of fern frond, page 115

Jill Britton: Fibonacci-series pinecone illustration, page 28

Fabio Cesari: Mandelbrot Set image, page 118

Les Enluminures, Paris and Chicago: Tally sticks, page 8

Noel Griffin, Spanky Fractal Database: Endpapers fractal image

Heinz Nixdorf MuseumsForum, Paderborn, Germany: Pascal's Calculating Machine, page 94

Paul Kremer: Periodic table, pages 78–79

M. S. Miesch (N.C.A.R., U.S.A.), A. S. Brun (C.E.A., Saclay, France), and J. Toomre (J.I.L.A., University of Colorado, U.S.A.): Image of convective patterns, page 117

NASA/WMAP: Images of planets, page 18; spiral galaxy, 27; cosmic microwave image of Milky Way, page 90; galaxy cluster, page 116

Science Photo Library: Computer simulation of jet aircraft taking off, page 105

All other illustrations courtesy of Richard Burgess.